Table of Contents

1 Chapter 1: The Essence of Voice Fidelity . . . 3
1.1 Understanding Signal Purity 6
1.2 Sources of Audio Degradation 11

2 Chapter 2: Antennas and Amplification 15
2.1 Antenna Fundamentals 19
2.2 Amplification Techniques 23
2.3 Minimizing Noise in Amplifiers 26

3 Chapter 3: Filtering the Noise Floor 31
3.1 Basic Filter Designs 35
3.2 Advanced Filtering Methods 40
3.3 Adaptive Noise Cancellation 43
3.4 Impulse Noise Reduction 48
3.5 Filter Implementation Strategies 53

4 Chapter 4: Modulation and Demodulation . . . 59
4.1 Common Modulation Schemes 63
4.2 Demodulation Challenges 67
4.3 Optimizing Demodulation 72

5 Chapter 5: Encoding for Clarity 77
5.1 Voice Encoding Basics 81

5.2 Advanced Voice Codecs 85
5.3 Error Correction Techniques 89

6 Chapter 6: Digital Signal Processing 95
6.1 DSP Fundamentals 100
6.2 Echo Cancellation Algorithms 103
6.3 Speech Enhancement Techniques 107

7 Chapter 7: Practical Implementation & Testing 113
7.1 Hardware Considerations 117
7.2 Software Tools & Libraries 122
7.3 Testing and Validation 127
7.4 Troubleshooting Common Issues 131
7.5 Real-World Case Studies 135

1 Chapter 1: The Essence of Voice Fidelity

The human voice, a delicate instrument of communication, conveys not just words but also emotion, nuance, and identity. Achieving high voice fidelity, therefore, goes beyond mere audibility; it demands preserving the integrity of the original sound wave as it travels from source to receiver. Understanding this underlying principle is the crucial first step in optimizing any voice communication system. Without a firm grasp of what constitutes "good" voice fidelity, efforts to improve signal reception become aimless and potentially counterproductive. Think of it as trying to restore a damaged painting without first understanding the artist's original vision. We aim not just to hear, but to truly understand.

The concept of signal purity is paramount. Imagine a pristine stream flowing down a mountainside. This represents the ideal voice signal – clear, unadulterated, and free from any contaminants. But as the stream flows further, it picks up sediment, pollution, and other impurities. Similarly, a voice signal, as it propagates through the air or travels through electronic circuits, inevitably accumulates noise and distortions. A pure signal is one where the original waveform is faithfully represented with minimal addition of extraneous elements. This means ensuring that the frequency response is flat across the audible spectrum, that harmonic distortion is kept to a minimum, and that the signal-to-noise ratio is as high as possible. Achieving this requires careful attention to every stage of the transmission and reception process, from the microphone to the speaker or headphones. It isn't simply about having loud and clear audio; it's about the absence of unwanted artifacts that color or obscure the original sound. Now, let's dissect the primary offenders that compromise audio fidelity. Environmental noise is the most obvious culprit. This encompasses a wide range of sounds, from the hum of electrical equipment to the rumble of traffic to the chatter of nearby conversations. These sounds can mask the voice signal, making it difficult to understand. But environmental noise is only one piece of the puzzle. Electrical interference, such as radio frequency interference (RFI) and electromagnetic interference (EMI), can also introduce unwanted noise into the system. These sources can emanate from power lines, electronic devices, and even the components within the communication equipment itself. Then there's the issue of hardware limitations. Microphones, amplifiers, and speakers all have their own inherent limitations in terms of frequency response, dynamic range, and distortion. A cheap microphone, for instance, might have a limited frequency response, meaning that it cannot accurately capture the full range of frequencies present in the human voice. Similarly, a poorly designed amplifier can introduce harmonic distortion, adding unwanted overtones to the signal. Finally, signal processing artifacts can degrade voice fidelity. Compression algorithms, while useful for reducing bandwidth, can also introduce distortion and reduce the dynamic range of the signal. Echo cancellation algorithms, while essential for improving the clarity of two-way communication, can sometimes introduce artifacts if not properly implemented. It's a layered challenge, but understanding the enemy is the first step toward victory.

Consider a scenario: you are attempting to record a podcast in your home office. The room isn't acoustically treated, so you are dealing with echoes and reverberations. Your computer fan is whirring, adding a constant hum to the background. A nearby radio transmitter is leaking RFI into your audio interface. And to top it off, the microphone you are using is a budget model with a limited frequency response. The result is a recording that sounds muddy, noisy, and unnatural. The first step is recognizing the sources of the problems. The echo from the room suggests the need for sound dampening. The computer fan noise means repositioning the computer or finding ways to quiet it. The RFI points to proper shielding of your audio cables, and a better microphone will definitely help. This diagnosis is more than half the battle.

The key takeaway here is that achieving high voice fidelity is a multifaceted challenge that requires a deep understanding of the factors that contribute to signal degradation. By identifying and addressing these factors, we can create voice communication systems that deliver clear, natural, and intelligible audio. This foundation, the understanding of the signal itself and the challenges it faces, is what empowers the subsequent chapters and their technical explorations. This careful approach to understanding the essentials of voice clarity lays the groundwork for deeper technical explorations in the book and will ultimately help anyone build voice communication systems that deliver crystal-clear audio.

1.1 Understanding Signal Purity

Signal purity isn't just about having a loud, clear voice; it's about having a signal devoid of unwanted artifacts that can mask or distort the intended message. Imagine trying to read an ancient scroll where the words are faded and smudged – that's what a compromised audio signal sounds like. We're not just aiming for audibility, but for absolute intelligibility, where the nuances and inflections of the voice are preserved. Think of it as restoring that ancient scroll, carefully cleaning away the blemishes to reveal the original script in all its glory. Achieving this requires a deep understanding of what constitutes a "pure" signal and how to protect it from the myriad forces that conspire to corrupt it. It's the foundation upon which all other optimizations rest, and without it, the most sophisticated antennas, amplifiers, and filters will struggle to deliver truly exceptional voice fidelity.
What degrades this purity? The sources are numerous and pervasive, ranging from the immediate environment to the very components we use to capture and transmit the voice.

Ambient noise – the cacophony of everyday life – is an obvious culprit. Street noise bleeding into a microphone, the hum of air conditioning, or even the subtle rustling of clothing can all contaminate the audio signal. Electrical interference is another significant factor. Radiated emissions from nearby devices, power lines, or even poorly shielded cables can introduce unwanted frequencies into the signal path. This can manifest as static, hum, or even more complex distortions that can be incredibly difficult to track down. Then, there are the inherent limitations of the hardware itself. Microphones, for instance, have their own noise floors and frequency response characteristics that can subtly alter the character of the voice. Amplifiers, while boosting the signal, can also introduce their own distortions and noise. Even the simple act of digitizing the audio signal can introduce quantization errors, which, if not carefully managed, can audibly degrade the sound. Finally, the transmission medium itself can be a source of degradation. Radio waves can be affected by atmospheric conditions, physical obstructions, and multi-path interference, where the signal reflects off surfaces and arrives at the receiver at slightly different times, creating echoes and distortions. Each of these factors contributes to a loss of signal purity, and mitigating their effects is crucial for achieving optimal voice fidelity.

Let's delve into practical ways to safeguard signal purity, starting with the most basic, yet often overlooked, aspect: the environment. Choosing a quiet recording or speaking location is paramount. If you can't completely eliminate ambient noise, consider using acoustic treatment to absorb sound reflections and reduce reverberation. Simple measures like closing windows, using rugs or carpets, and adding acoustic panels to walls can make a significant difference. Microphones are your gateway to capturing the voice signal, so selecting the right one is crucial. Consider the microphone's polar pattern – its sensitivity to sound from different directions. A cardioid microphone, for example, is most sensitive to sound directly in front of it and rejects sound from the sides and rear, making it ideal for reducing background noise. Dynamic microphones are generally more robust and less susceptible to noise than condenser microphones, but condenser microphones often offer a wider frequency response and greater sensitivity, which can be beneficial in quieter environments. Experiment with different microphone types and positions to find the optimal setup for your specific needs. Cable quality matters! Using shielded cables can significantly reduce the amount of electrical interference that enters the signal path. Ensure that all connections are secure and free from corrosion, as loose connections can introduce noise and signal loss. When using wireless microphones or communication systems, be mindful of frequency selection. Choose frequencies that are less likely to be affected by interference from other devices or services. Regularly scan the spectrum to identify and avoid congested frequencies. Implementing these environmental and hardware-related precautions provides a solid foundation for maintaining signal purity.

Next, let's examine advanced techniques for addressing more complex sources of signal degradation. One powerful approach is to employ digital signal processing (DSP) techniques to remove noise and enhance the clarity of the voice signal. Noise reduction algorithms can analyze the audio and identify and suppress unwanted background noise. Spectral subtraction is one such technique, which estimates the noise floor and subtracts it from the signal, effectively removing the noise while preserving the voice. Another important DSP technique is equalization, which allows you to adjust the frequency balance of the audio signal to compensate for deficiencies in the microphone, amplifier, or transmission medium. You can use equalization to boost frequencies that are lacking or attenuate frequencies that are overly prominent, resulting in a more balanced and natural-sounding voice. Echo cancellation is essential for two-way communication, as it prevents feedback and ensures clear and intelligible conversations. Echo cancellation algorithms analyze the incoming audio signal and remove any echoes that are present, allowing for seamless and natural communication. In modern communication systems, advanced codecs play a crucial role in preserving signal purity while minimizing bandwidth requirements. These codecs use sophisticated compression techniques to encode the voice signal efficiently, while also incorporating error correction mechanisms to mitigate the effects of data loss or corruption during transmission. Choosing the right codec for your specific application can significantly improve voice fidelity, especially in challeng-

ing network conditions.
However, technology alone isn't the answer. Careful monitoring and analysis of the audio signal are essential for identifying and addressing any remaining issues. Use headphones or studio monitors to critically listen to the audio and identify any unwanted noise, distortion, or other artifacts. Spectrum analyzers can be used to visualize the frequency content of the audio signal, allowing you to identify specific sources of interference or distortion. Regularly test and validate your voice communication system to ensure that it meets your desired standards of fidelity. This includes conducting listening tests, measuring signal-to-noise ratios, and evaluating the overall clarity and intelligibility of the voice signal. Document your findings and make adjustments as necessary to optimize performance.
Finally, remember that maintaining signal purity is an ongoing process, not a one-time fix. The environment, hardware, and network conditions can change over time, so it's important to regularly monitor and adapt your strategies accordingly. Stay up-to-date on the latest technologies and techniques for voice fidelity optimization, and be willing to experiment and innovate to find new and better ways to achieve optimal results. The pursuit of signal purity is a journey, not a destination, and the rewards – clear, intelligible, and natural-sounding voice communication – are well worth the effort.

1.2 Sources of Audio Degradation

Audio degradation, an insidious foe of clear voice communication, manifests in a multitude of ways, often subtly undermining the integrity of a received signal. Understanding the root causes of these degradations is the first crucial step toward mitigating their effects and achieving optimal voice fidelity. This chapter will delve into these sources, dissecting their mechanisms and laying the groundwork for the signal enhancement strategies discussed in subsequent sections.

One of the most pervasive sources of audio degradation is ambient noise. This encompasses a broad spectrum of unwanted sounds present in the environment surrounding both the transmitter and the receiver. Consider a crowded office where background chatter, keyboard clicks, and the hum of air conditioning all contribute to a cacophony that interferes with the clarity of a voice transmission. Likewise, imagine the roar of traffic or the blare of sirens bleeding into a conversation conducted outdoors. Such extraneous sounds add unwanted frequencies to the desired signal, effectively masking portions of the voice and making it difficult to discern individual words or phrases. The severity of ambient noise depends on factors such as the intensity of the noise sources, their proximity to the microphone, and the acoustic properties of the environment. Reflective surfaces, for instance, can amplify noise levels by creating echoes and reverberations, further compounding the issue. This type of noise is not limited to just environmental sounds, but also the intrinsic equipment's own produced frequencies.

Beyond ambient noise, interference from other electronic devices represents another significant source of degradation. Radio frequency interference (RFI), in particular, can wreak havoc on voice signals transmitted wirelessly. This interference arises when electromagnetic waves emitted by devices such as mobile phones, Wi-Fi routers, and microwave ovens overlap with the frequencies used for voice communication. The resulting interference can manifest as static, buzzing, or even complete signal dropouts, rendering the voice transmission unintelligible. Consider the scenario where a Bluetooth headset experiences interference from a nearby Wi-Fi router operating on the same frequency band. The router's signal can overpower the headset's signal, causing disruptions in the audio stream and compromising voice clarity. Shielding of electronic circuits and cables is often insufficient to fully protect against the adverse effects of electromagnetic waves. It is therefore essential to plan equipment layout carefully to minimize the potential for electromagnetic interference.

Signal attenuation, the gradual loss of signal strength as it travels from transmitter to receiver, also constitutes a major contributor to audio degradation. This phenomenon is particularly pronounced in wireless communication systems, where signals are subject to free-space path loss. As the distance between the transmitter and receiver increases, the signal strength decreases exponentially, making it more susceptible to noise and interference. Obstacles in the signal path, such as walls, buildings, and foliage, can further attenuate the signal, creating "dead spots" where voice communication is severely impaired. The effects of signal attenuation can be mitigated by increasing the transmit power, using directional antennas to focus the signal, or deploying repeaters to amplify the signal along its path. However, these measures must be carefully implemented to avoid introducing additional sources of distortion or interference. It is important to also acknowledge that environmental conditions, such as humidity and temperature, can also affect the severity of attenuation.

Hardware limitations within the audio equipment itself can also degrade voice fidelity. Microphones, for example, may exhibit non-linear frequency responses, meaning they do not accurately capture all frequencies within the human voice range. This can lead to a distorted or unnatural sound. Speakers, similarly, may introduce distortion, particularly at high volume levels, or may have an uneven frequency response that emphasizes certain frequencies over others. Amplifiers, too, can contribute to audio degradation if they are not properly designed or calibrated. Distortion, noise, and limited bandwidth are all potential issues that can arise from poorly implemented amplification circuitry. Regular maintenance and calibration are important for ensuring consistent performance.

Finally, quantization noise, a form of distortion inherent in digital audio systems, can also degrade voice fidelity. Quantization noise arises when an analog voice signal is converted into a digital representation. The analog signal, which is continuous, must be sampled and converted into discrete values. The accuracy of this conversion depends on the number of quantization levels used. If the number of levels is insufficient, the resulting digital signal will be a coarse approximation of the original analog signal, leading to audible distortion, particularly during quiet passages. The degree of quantization noise is directly related to the bit depth of the audio signal. Higher bit depths provide more quantization levels, resulting in a more accurate representation of the analog signal and reduced quantization noise. Over-sampling and dithering are two techniques commonly employed to minimize the audibility of quantization noise.

By understanding the sources of audio degradation – ambient noise, interference, signal attenuation, hardware limitations, and quantization noise – we can begin to develop strategies for mitigating their effects. The following chapters will explore a range of techniques, from noise reduction algorithms to advanced filtering methods, that can be used to optimize signal reception and achieve crystal-clear voice fidelity. The journey to pristine audio begins with a deep understanding of its potential pitfalls.

2 Chapter 2: Antennas and Amplification

Antennas, acting as the initial gateway for incoming signals, deserve careful consideration. They're not just hunks of metal; they are intricately designed components that dictate the strength and clarity of the signal you receive. A dipole antenna, for instance, a simple and commonly used design, consists of two conductive elements, typically rods or wires, extending outwards from a central feed point. Its length is directly related to the wavelength of the signal it's designed to receive; a half-wavelength dipole is most efficient at receiving signals at its resonant frequency. However, the world isn't just about theoretical ideal scenarios. Impedance matching becomes crucial. An antenna's impedance, a measure of its opposition to the flow of alternating current, must match the impedance of the receiving equipment (usually 50 ohms). Mismatches lead to signal reflections, reducing the amount of power transferred to the receiver and potentially causing signal distortion. Antenna placement also drastically influences performance. A clear line of sight to the transmitting source is paramount. Obstructions like buildings, trees, and even terrain can attenuate signals, causing them to weaken or be reflected away. Multipath interference, where the signal arrives at the receiver via multiple paths, can also cause distortion and fading. Experimenting with different antenna locations and orientations can significantly improve signal reception.

Amplification, the process of increasing the power of a signal, is often necessary to boost weak signals received by the antenna to a level suitable for processing. Operational amplifiers, or op-amps, are versatile building blocks commonly used in audio amplification circuits. They offer high gain, low noise, and can be configured in various ways to achieve different amplification characteristics. A non-inverting amplifier configuration, for example, provides a stable and predictable gain, while an inverting amplifier configuration offers greater flexibility in impedance matching. Gain isn't everything, though. Amplifier linearity is crucial for preserving the integrity of the voice signal. A linear amplifier amplifies the signal without introducing distortion, meaning that the output signal is a faithful representation of the input signal, only larger. Non-linear amplifiers, on the other hand, can introduce harmonic distortion, creating unwanted frequencies that were not present in the original signal. This can manifest as a muddy or unclear sound. The choice of amplifier depends on the specific application and the desired performance characteristics. For critical applications where voice fidelity is paramount, Class A amplifiers, known for their excellent linearity, might be preferred, despite their lower efficiency compared to Class B or Class D amplifiers.

Noise, an ever-present enemy of signal clarity, contaminates the desired signal, masking subtle details and reducing intelligibility. The noise figure (NF) of an amplifier quantifies the amount of noise it adds to the signal. A lower noise figure indicates better performance. Thermal noise, also known as Johnson noise, is inherent in all electronic components due to the random motion of electrons. It's directly proportional to temperature and resistance, meaning that components operating at higher temperatures and with higher resistances will generate more thermal noise. Shot noise arises from the discrete nature of electric charge and is particularly prevalent in semiconductor devices like transistors and diodes. Flicker noise, also known as 1/f noise, is a type of noise whose power spectral density is inversely proportional to frequency. It's more prominent at lower frequencies and can be a significant source of noise in audio applications. Careful component selection, choosing low-noise resistors and transistors, can minimize noise generation. Biasing the amplifier circuit for optimal noise performance is also crucial. Shielding the amplifier circuit from external sources of interference, such as electromagnetic radiation from other electronic devices, can further reduce noise. This can be achieved by enclosing the amplifier circuit in a metal enclosure.

To minimize noise, consider these strategies: input impedance is crucial, matching the amplifier's input impedance to the source impedance minimizes signal reflections and maximizes power transfer, reducing the potential for noise to be amplified along with the desired signal. Employing feedback networks strategically stabilizes the amplifier's gain and reduces distortion, but also lowers the noise. A well-designed feedback network improves the amplifier's overall performance and reduces the amount of noise that's amplified. Bandwidth limitations in amplifiers can inadvertently amplify noise beyond the required frequency range, so restricting the amplifier's bandwidth to the relevant audio frequencies reduces the overall noise level without affecting the desired signal. Implementing a pre-amplifier stage ahead of the main amplifier boosts the signal level before it encounters the main amplifier's noise, thereby improving the signal-to-noise ratio (SNR). This effectively reduces the relative impact of the main amplifier's noise on the overall signal quality. Finally, cooling the amplifier components reduces thermal noise, particularly in sensitive low-noise amplifiers (LNAs) used in high-fidelity applications. This strategy is more common in specialized applications where extremely low noise is required.

In conclusion, achieving optimal voice fidelity requires careful attention to antennas and amplification. Selecting the right antenna, ensuring proper impedance matching, and optimizing antenna placement are crucial for capturing a strong and clear signal. Choosing linear amplifiers with low noise figures and implementing effective noise reduction techniques are essential for preserving the integrity of the voice signal during amplification. By understanding these principles and applying them diligently, you can significantly improve the clarity and intelligibility of voice communication systems. The pursuit of "Parchment's Whispers" begins here, with a commitment to capturing and amplifying every nuance of the human voice with the utmost precision.

2.1 Antenna Fundamentals

Antennas are the unsung heroes of voice communication, the initial point of contact for the faint whispers traveling through the air. Without a properly functioning antenna, even the most sophisticated signal processing techniques will fall flat. In essence, the antenna acts as a transducer, converting electromagnetic waves into electrical signals that your receiver can then interpret. Choosing the right antenna, understanding its characteristics, and properly implementing it are paramount to achieving the voice fidelity we're striving for. Therefore, we need to understand the characteristics to find the one that adapts better to our needs.

The world of antennas is vast and varied, offering a plethora of options each with its own strengths and weaknesses. For voice communication systems, particularly those focused on clarity, a few key antenna types stand out. Dipole antennas, perhaps the simplest and most fundamental, consist of two conductive elements, typically rods or wires, connected to a feed point. Their omnidirectional radiation pattern makes them relatively easy to position, but they can be susceptible to interference from all directions. Yagi-Uda antennas, recognizable by their array of parallel elements, offer significantly higher gain and directivity. This means they can focus their reception in a specific direction, improving signal strength and reducing interference, but they require careful aiming. Loop antennas, often used in mobile devices, are compact and less sensitive to polarization changes, making them a good choice for situations where the orientation of the transmitting antenna is unknown. Finally, patch antennas, commonly found in wireless communication systems, are low-profile and can be easily integrated into devices. Understanding these basic types and their respective trade-offs is the first step in selecting the optimal antenna for your specific application. It is also very important to keep in mind the frequency of the voice signal.

Beyond the basic types, several key characteristics define an antenna's performance. Gain, measured in decibels (dBi), indicates how much the antenna amplifies the signal in a specific direction compared to an isotropic radiator (a theoretical antenna that radiates equally in all directions). A higher gain antenna will generally provide a stronger signal, but at the cost of a narrower beamwidth. Beamwidth, measured in degrees, describes the angle over which the antenna's radiation pattern is concentrated. A narrow beamwidth provides higher gain and better interference rejection, but requires more precise aiming. Impedance matching is crucial for efficient signal transfer between the antenna and the receiver. Mismatched impedance can lead to signal reflections and loss of power. Antennas are designed to operate at a specific impedance, typically 50 ohms, and it's important to ensure that the receiver's input impedance matches the antenna's impedance. VSWR (Voltage Standing Wave Ratio) is a measure of how well the antenna's impedance is matched to the transmission line. A VSWR of 1:1 indicates a perfect match, while higher values indicate increasing levels of mismatch and signal loss. Also, polarity is important, it defines the orientation of the electric field.

Selecting the right antenna involves considering several factors, including the frequency of the voice signal, the desired range, the environment, and the potential for interference. For longer-range communication, a high-gain directional antenna like a Yagi-Uda might be necessary. In urban environments with high levels of interference, a directional antenna can help to focus on the desired signal and reject unwanted noise. For mobile applications, a compact antenna like a loop or patch antenna is often preferred. The polarization of the transmitting antenna should also be considered. If the transmitting antenna is vertically polarized, the receiving antenna should also be vertically polarized for optimal signal reception. If you are unsure about the transmitting antenna's polarization, a circularly polarized antenna can be a good choice, as it can receive signals of any polarization. In this case, it may be better to focus on an antenna that helps you to filter noises that alter the voice signal.

Beyond selecting the right antenna, proper installation and maintenance are crucial for optimal performance. Ensure that the antenna is securely mounted and properly grounded to prevent damage from lightning and static electricity. Use high-quality coaxial cables to connect the antenna to the receiver, and keep the cable runs as short as possible to minimize signal loss. Regularly inspect the antenna for damage, such as corrosion or broken elements. Over time, exposure to the elements can degrade the antenna's performance. Clean the antenna periodically to remove dirt and debris, which can also affect signal reception. A poorly installed antenna can significantly degrade signal quality, even if it is the best antenna for the job. Make sure all cables and connectors are tight and weatherproofed. This is a practice that everyone should take care to avoid problems.

Finally, optimizing antenna placement can significantly improve signal reception. Experiment with different locations and orientations to find the sweet spot where the signal is strongest and the interference is minimal. Avoid placing the antenna near obstructions, such as buildings or trees, which can block or reflect the signal. Consider using a signal meter to measure the signal strength at different locations. This can help you to identify the optimal placement for your antenna. In some cases, it may be necessary to use multiple antennas in a diversity configuration to improve signal reception. A diversity system uses two or more antennas to receive the same signal, and then combines the signals to improve the overall signal quality. This can be particularly effective in environments with multipath fading, where the signal is reflected off of multiple objects, causing interference. If this is not possible, consider using a signal amplifier to boost the signal strength. However, be aware that amplifiers can also amplify noise, so it's important to use a high-quality ampli-

fier with a low noise figure. In this way, optimizing antenna placement can significantly improve signal reception.
Remember, the antenna is the first line of defense in the battle for voice fidelity. By understanding the fundamentals of antenna operation, selecting the right antenna for your application, and properly installing and maintaining it, you can significantly improve the clarity and reliability of your voice communication system. It is also important to remember that every voice signal is different, so it is important to experiment with different antenna configurations to find what works best for you. It is a task that requires attention to the smallest details.
Think of the antenna as your ear, straining to hear a distant whisper. Just as you might cup your hand behind your ear to focus the sound, the antenna focuses electromagnetic energy. The better the antenna, the clearer the whisper becomes.

2.2 Amplification Techniques

Amplification, at its core, is about taking a weak signal and making it stronger. In the context of voice fidelity, this means boosting the electrical signal representing the spoken word without adding distortion or, crucially, amplifying any accompanying noise. The art, and the science, lies in achieving that delicate balance. We aren't just aiming for a louder signal; we're aiming for a clearer one. A crucial first step is understanding your current system. Where is the weakest link? Is it the microphone pre-amplifier, the distance of transmission, or a combination? Once you identify the bottleneck, you can strategically apply amplification techniques. Let's delve into some proven strategies and considerations.

One of the simplest, yet often overlooked, techniques is optimizing the antenna. This isn't always about raw amplification, but about ensuring you're capturing the strongest possible initial signal. If you're using a directional antenna, careful alignment is paramount. A slight misalignment can drastically reduce signal strength. Consider experimenting with different antenna types to see which performs best in your particular environment. A Yagi-Uda antenna, for example, offers high gain in a specific direction, ideal for point-to-point communication, while a dipole antenna provides a wider coverage pattern. Furthermore, the placement of the antenna is critical. Keep it as far away from sources of interference as possible, such as power lines, electronic devices, and other antennas. Even seemingly innocuous objects can reflect or absorb radio waves, so experiment with different locations to find the sweet spot. Grounding the antenna properly is also essential for reducing noise and improving signal quality. A well-grounded antenna acts as a shield, diverting unwanted currents away from the receiver.

Next, we can consider pre-amplifiers. These devices are designed to boost the signal early in the chain, ideally right after the antenna, before it has a chance to degrade significantly. The key here is to select a pre-amplifier with a low noise figure. The noise figure is a measure of how much noise the amplifier adds to the signal. A lower noise figure is always better.

Also, pay attention to the gain of the pre-amplifier. Too much gain can overload the subsequent stages of the receiver, leading to distortion. A gain that is too low won't provide sufficient amplification. A variable gain pre-amplifier allows you to fine-tune the amplification to match the signal strength and the noise level. Remember to shield the pre-amplifier from external interference. A metal enclosure can effectively block unwanted signals from entering the amplifier and corrupting the signal. And, crucially, ensure the pre-amplifier is properly powered. A stable power supply is essential for consistent performance. Fluctuations in the power supply can introduce noise and distortion into the signal.

Another critical factor is impedance matching. Maximum power transfer occurs when the impedance of the source (e.g., antenna) matches the impedance of the load (e.g., amplifier). If the impedances are mismatched, some of the signal will be reflected back to the source, reducing the amount of power that is delivered to the load. This can result in a weaker signal and increased noise. Use impedance matching networks, such as L-networks or transformers, to ensure that the impedances are properly matched. These networks can be designed to match a wide range of impedances, ensuring optimal signal transfer.

Moving beyond hardware, consider the cables you are using. Low-loss cables are essential for minimizing signal attenuation, especially over long distances. Coaxial cables are commonly used for transmitting radio frequency signals. Choose a cable with a low attenuation rating, such as LMR-400 or

RG-8. Also, ensure that the connectors are properly installed. Loose or corroded connectors can introduce signal loss and noise. Use high-quality connectors and crimp them properly using the appropriate tools. Weatherproofing the connectors is also important, especially in outdoor environments. Moisture and corrosion can significantly degrade the performance of the connectors.

Lastly, proper grounding and shielding cannot be overstated. Ground loops can introduce significant noise into the system. Ensure that all equipment is properly grounded to a common ground point. Use shielded cables to prevent external interference from entering the signal path. The shielding should be properly grounded to the chassis of the equipment. Furthermore, consider using ferrite beads on cables to suppress high-frequency noise. Ferrite beads act as inductors, blocking high-frequency currents while allowing low-frequency currents to pass through. They can be particularly effective at reducing noise from power supplies and digital circuits. By implementing these techniques, you can significantly improve the signal-to-noise ratio and enhance the clarity of voice signals. The quest for perfect voice fidelity is a journey, not a destination. Continual experimentation and refinement are key.

2.3 Minimizing Noise in Amplifiers

Noise, an ever-present adversary in the realm of amplification, can severely compromise voice fidelity. It creeps into the system at every stage, from the initial antenna pickup to the final speaker output, and its insidious effects manifest as unwanted hiss, crackle, and distortion, obscuring the subtle nuances of the human voice. For the novice electronics enthusiast, understanding and mitigating noise is paramount to achieving clear, intelligible audio. One must remember that the amplifier, while boosting the desired signal, also amplifies any noise present at its input. Therefore, a seemingly minor source of noise can be magnified to unacceptable levels.

The first line of defense against noise lies in careful component selection. Resistors, often overlooked, can be significant contributors to thermal noise, also known as Johnson-Nyquist noise. This noise is inherent to all resistive materials and increases with temperature. When selecting resistors, particularly in the input stages of an amplifier, opt for metal film resistors over carbon composition resistors. Metal film resistors exhibit lower thermal noise and are generally more stable. Furthermore, consider the resistor's value; higher resistance values tend to generate more thermal noise. Therefore, when designing the amplifier circuit, strive to use the lowest practical resistance values that still meet the circuit's functional requirements. The specific type of amplifier being used is also incredibly important. Op-amps, for example, have input noise voltage and current specifications that can vary widely between different models. Always consult the datasheet and select an op-amp with low input noise characteristics, especially for the initial amplification stages. Similarly, transistors, whether BJTs or FETs, are sources of noise. BJTs exhibit shot noise and thermal noise, while FETs primarily exhibit thermal noise and flicker noise. The choice of transistor type and its operating point can significantly impact the overall noise performance of the amplifier.

Circuit layout and grounding techniques are equally crucial in minimizing noise. Poorly designed circuit boards can introduce ground loops, which are closed conductive paths that act as antennas, picking up electromagnetic interference (EMI) from the surrounding environment. To avoid ground loops, employ a star grounding scheme, where all ground connections converge at a single point. This minimizes the circulating currents that cause noise. Furthermore, keep signal traces as short as possible and away from potential noise sources, such as power supplies and digital circuits. Shielding is another effective technique for reducing noise pickup. Enclosing the amplifier circuit in a metal enclosure or using shielded cables can block external electromagnetic fields from interfering with the signal. Pay particular attention to the input stage, as it is most susceptible to noise. Additionally, consider using bypass capacitors strategically placed throughout the circuit. These capacitors provide a low-impedance path for high-frequency noise to ground, preventing it from propagating through the circuit.

Power supply noise can also wreak havoc on voice fidelity. Ripple and noise from the power supply can directly modulate the amplified signal, resulting in unwanted hum and distortion. To mitigate power supply noise, use a well-regulated power supply with low ripple and noise specifications. Decoupling capacitors placed near the amplifier's power supply pins can further reduce noise. These capacitors act as local energy reservoirs, providing a stable voltage source and filtering out high-frequency noise. Remember that the quality of the power supply directly impacts the quality of the amplified signal.

Beyond these foundational techniques, more advanced methods can be employed to further minimize noise. One such technique is noise cancellation. Differential amplifiers, for example, are inherently good at rejecting common-mode noise, which is noise that is present on both input terminals. By carefully designing the amplifier circuit to be symmetrical and balanced, common-mode noise can be effectively cancelled out. Another advanced technique is the use of low-noise amplifiers (LNAs). These specialized amplifiers are designed with extremely low noise figures and are often used in the front-end stages of sensitive receivers. They can provide a significant improvement in signal-to-noise ratio, but they also tend to be more expensive and require careful design considerations.

Finally, don't underestimate the importance of proper testing and troubleshooting. After building the amplifier, thoroughly test its noise performance using an oscilloscope or spectrum analyzer. Measure the noise floor and identify any specific noise frequencies that may be present. If excessive noise is detected, systematically troubleshoot the circuit to identify the source. This may involve probing different points in the circuit with an oscilloscope, checking component values, and verifying grounding connections. Remember that noise reduction is an iterative process, and it may take some experimentation and fine-tuning to achieve optimal results. While it is impossible to completely eliminate noise, by understanding its sources and employing effective mitigation techniques, you can significantly improve the fidelity and clarity of amplified voice signals, allowing the subtle nuances of the human voice to shine through. Don't be afraid to experiment with different techniques and components, as the optimal solution will often depend on the specific application and the characteristics of the noise environment. Always double-check your work; a single overlooked connection or a poorly chosen component can undo hours of careful design.

3 Chapter 3: Filtering the Noise Floor

Imagine a concert hall. The orchestra's tuning, the murmur of the audience, the distant rumble of city traffic – all blend into a constant, pervasive hum. This is the noise floor, the base level of unwanted sound that obscures the music we came to hear. In voice communication, this "noise floor" represents the background interference that diminishes the clarity of the intended voice signal. To truly optimize voice fidelity, we must meticulously filter out this noise.

Filtering, at its most basic, is about selectively passing certain frequencies while blocking others. A simple low-pass filter, for example, allows low-frequency components to pass through while attenuating high frequencies. Imagine a sieve used to separate flour from larger clumps; the sieve allows the fine flour to pass but retains the clumps. Similarly, a high-pass filter does the opposite, allowing high frequencies to pass while blocking low frequencies. A band-pass filter, as the name suggests, allows only a specific band of frequencies to pass, blocking both higher and lower frequencies. These filters can be constructed using basic electronic components like resistors, capacitors, and inductors. Each component interacts with the others to create a frequency-dependent impedance, effectively attenuating unwanted frequencies. The design of these filters involves choosing the right component values to achieve the desired cutoff frequency and attenuation characteristics. For instance, in a simple RC low-pass filter, the cutoff frequency (the frequency at which the signal is attenuated by 3dB) is inversely proportional to the product of the resistance and capacitance. So, increasing either the resistance or the capacitance will lower the cutoff frequency.

However, the real world is rarely so simple. Static filters, while useful, often fall short in dynamic environments where the noise characteristics change over time. That's where advanced filtering methods come into play. Adaptive filters, for instance, automatically adjust their characteristics based on the incoming signal. Think of a chameleon that changes its color to blend in with its surroundings; an adaptive filter constantly "learns" the noise characteristics and adjusts its filtering parameters to minimize its impact. This is achieved through algorithms that analyze the input signal and modify the filter coefficients to minimize the error between the desired signal and the filtered output. One common type of adaptive filter is the Least Mean Squares (LMS) filter, which iteratively adjusts the filter coefficients to minimize the mean square error. The LMS algorithm is relatively simple to implement and can be effective in reducing noise in many applications. Another powerful adaptive filtering technique is the Kalman filter, which is particularly useful when dealing with noisy signals and uncertain system dynamics. The Kalman filter uses a state-space model to estimate the signal and noise characteristics and then uses this information to optimally filter the signal. These filters require more computational power but can provide superior performance in challenging environments.

Adaptive noise cancellation takes filtering a step further by actively subtracting the noise from the signal. The fundamental principle involves using a reference microphone to capture the ambient noise. This noise signal is then processed and subtracted from the primary signal, which contains both the desired voice and the noise. The key challenge lies in accurately estimating the noise component in the primary signal. This often involves using adaptive algorithms to minimize the correlation between the noise reference and the desired signal. Imagine two waves, one representing the noise and the other an "anti-noise" wave that cancels it out. Achieving perfect cancellation is rarely possible, but even a partial reduction in noise can significantly improve voice clarity. Practical implementations often involve sophisticated signal processing techniques to account for factors like acoustic echo and variations in the noise environment.

Sometimes, the noise isn't a constant hum, but rather sharp, transient bursts – impulse noise. Think of the crackle of static electricity or the click of a keyboard. These brief but intense disturbances can be particularly jarring and distracting. Impulse noise reduction techniques aim to identify and suppress these transient events without affecting the underlying voice signal. One common approach involves using a threshold detector to identify impulses based on their amplitude or duration. Once an impulse is detected, it can be suppressed by clipping the signal or replacing it with an interpolated value. More sophisticated techniques involve using wavelet transforms to analyze the signal in both time and frequency domains, allowing for more precise detection and removal of impulses. The goal is to minimize the impact of these transient noises on the perceived voice quality.

Ultimately, the effectiveness of any filtering strategy depends on how it's implemented. Filters can be implemented in hardware using analog or digital circuits, or in software using digital signal processing (DSP) algorithms. Hardware implementations offer the advantage of real-time processing and low latency, making them suitable for applications where immediate response is critical. Software implementations, on the other hand, offer greater flexibility and programmability, allowing for more complex filtering algorithms to be implemented. When designing a filter implementation, it's crucial to consider factors like computational complexity, power consumption, and cost. The choice of implementation will depend on the specific application requirements and the available resources. In a low-power embedded system, for example, a simple hardware filter might be the most practical solution, while in a high-performance communication system, a sophisticated DSP-based filter might be preferred. Careful consideration of these factors will ensure that the filtering

strategy is both effective and efficient. Choosing the right filter involves a careful balancing act: minimizing noise while preserving the integrity of the desired voice signal.
My own experience has taught me that there is no one-size-fits-all solution. The optimal approach depends on the specific characteristics of the noise and the desired voice signal. Experimentation and careful analysis are often necessary to fine-tune the filtering parameters and achieve the best possible results. Just as a skilled musician carefully adjusts the settings on their amplifier to achieve the perfect tone, so too must we carefully tailor our filtering strategies to optimize voice fidelity. The journey to pristine voice clarity is a continuous process of learning, adapting, and refining our techniques. By understanding the fundamentals of filtering and embracing the power of advanced methods, we can effectively "filter the noise floor" and unlock the full potential of voice communication.

3.1 Basic Filter Designs

Basic Filter Designs
The world of signal processing often feels like an endless battle against unwanted noise. In the quest for optimal voice fidelity, filters emerge as indispensable tools, acting as gatekeepers that allow desired frequencies to pass while attenuating or rejecting undesirable ones. For those just starting, understanding the fundamental filter designs is paramount. This chapter serves as a practical guide, introducing the most common filter types and providing insights into their applica-

tion in voice signal enhancement.

Let's begin with the bedrock of filtering: the low-pass filter. Imagine a sieve designed to separate larger pebbles from fine sand. A low-pass filter operates similarly, allowing low-frequency components of a signal to pass through while attenuating high-frequency components. These filters are invaluable in removing high-pitched noise, such as hiss or static, that can mask the clarity of a voice signal. Their simplicity makes them a common starting point in noise reduction efforts. For instance, in a noisy recording environment, a low-pass filter can effectively smooth out the harsher, high-frequency distractions, bringing the lower-frequency voice into sharper focus. But remember, excessive filtering can result in a muffled or unnatural sound, so careful tuning is always crucial.

Now, consider the opposite: the high-pass filter. Envision a bouncer at a club, only admitting those above a certain height. The high-pass filter functions analogously, permitting high-frequency components to pass through while attenuating low-frequency ones. This type of filter proves beneficial in eliminating low-frequency rumble or hum, often caused by electrical interference or environmental vibrations. In voice recordings, high-pass filters can remove unwanted "boominess" or proximity effect, resulting in a clearer, more intelligible signal. You'll often see them deployed in situations where the voice recording suffered from low-end resonance, which can drastically improve the quality of the spoken word. Still, be cautious. Aggressively applying a high-pass filter could

remove important low-frequency components of the voice itself, giving it a thin or unnatural timbre.
Next, we'll explore the band-pass filter. Think of a race official stationed on the track, and only focused on the middle-distance runner. As for the band-pass filter, it allows a specific range of frequencies to pass through while attenuating frequencies both above and below this range. This filter is particularly useful when you need to isolate the frequencies most relevant to the voice signal and reject noise that falls outside that band. For example, if you know that the most crucial voice information resides between 300 Hz and 3 kHz, you can design a band-pass filter to isolate that range, effectively reducing both low-frequency rumble and high-frequency hiss. Be sure to choose your bandwidth carefully, as a narrow bandwidth could exclude important components of the voice signal, while a broad bandwidth may let in too much noise.
Lastly, examine the band-stop filter, also known as a notch filter. Imagine a specialized tool designed to remove a specific splinter from a piece of wood. The band-stop filter rejects a specific range of frequencies while allowing frequencies outside that range to pass through. These filters are incredibly useful when dealing with very specific, troublesome frequencies, such as those caused by electrical hum or interference. For example, if you notice a persistent 60 Hz hum in your audio signal, a notch filter precisely tuned to that frequency can effectively remove it without significantly affecting the rest of the voice signal. Precision is key here; inaccurate

tuning can inadvertently remove desirable frequencies, negatively impacting the overall quality of the recording.

When implementing these basic filter designs, several factors can influence the result. Filter order, expressed as the number of reactive components (capacitors and inductors in analog filters, or taps in digital filters), determines the steepness of the filter's attenuation slope. Higher-order filters offer steeper slopes, meaning they can attenuate frequencies more aggressively outside the passband, leading to more effective noise reduction. However, they also introduce more phase distortion, which can negatively impact the naturalness of the voice signal. For beginners, starting with lower-order filters is advisable, gradually increasing the order as needed while carefully monitoring the impact on the signal.

Another important consideration is filter type. Butterworth filters provide a flat response in the passband and a smooth roll-off in the stopband, making them a good general-purpose choice. Chebyshev filters offer a steeper roll-off but introduce ripples in the passband or stopband. Bessel filters provide a linear phase response, minimizing phase distortion, but have a less steep roll-off. For voice applications, Butterworth filters are often preferred for their balanced performance. But, when the linearity is important, it might be the case to use a Bessel filter, even if it has a less steep roll-off in the stopband. It's a matter of balancing those parameters.

Transitioning from analog to digital filters unveils a new landscape of possibilities. Digital filters, implemented using algorithms on a digital signal processor (DSP), offer greater flexibility and precision than their analog counterparts. Finite Impulse Response (FIR) filters are inherently stable and can be designed with linear phase response, making them suitable for preserving the time-domain characteristics of the voice signal. Infinite Impulse Response (IIR) filters, on the other hand, can achieve steeper roll-off characteristics with fewer coefficients than FIR filters, making them computationally more efficient. However, IIR filters can be unstable and exhibit non-linear phase response, which may introduce artifacts in the processed signal.

The process of choosing and applying filters is not simply a technical exercise, it's an art that requires practice and careful listening. Start with simple filters and gradually increase their complexity as your understanding grows. Be mindful of the trade-offs between noise reduction and signal degradation, and always prioritize preserving the naturalness and intelligibility of the voice. It may be beneficial to use software plugins to visualize the frequency spectrum. This will help you to see what components are being attenuated.

Experiment with different filter types and orders to discover what works best for your specific application. Remember, the goal is not to eliminate all noise, but to improve the clarity and intelligibility of the voice signal without sacrificing its natural quality. Filtering is a delicate balancing act, requiring both technical knowledge and a keen ear.

3.2 Advanced Filtering Methods

Now let's explore the realm of enhanced filtering techniques, moving beyond the basics to unlock superior voice clarity. Where simple filters like low-pass and high-pass offer broad-stroke noise reduction, advanced methods provide targeted precision. This chapter assumes a foundational understanding of filter design, so we will focus on implementations.
Adaptive filters are the chameleons of the filtering world. Unlike fixed filters with static characteristics, they dynamically adjust their parameters based on the incoming signal. Think of them as filters that learn to recognize and eliminate specific types of noise, even when that noise is constantly evolving. This is particularly useful in environments with fluctuating background noise, where a static filter would quickly become ineffective. Adaptive filters achieve this by continuously analyzing the input signal and adjusting their coefficients to minimize the error between the desired signal and the actual output. One common type is the Least Mean Squares (LMS) filter, which iteratively adjusts its coefficients to minimize the mean square error. Another powerful adaptive filter is the Recursive Least Squares (RLS) filter, which offers faster convergence than LMS but at the cost of increased computational complexity. Selecting between LMS and RLS depends on the specific application requirements, balancing performance against computational resources. Imagine a scenario where you are trying to isolate a speaker's voice in a crowded room, the noise is constantly shifting as people move and talk, an adaptive filter can compensate for these changes in

real time, effectively focusing in on the desired voice.

Kalman filters represent another sophisticated filtering approach, going beyond simple noise reduction to provide optimal state estimation. Originating from control systems theory, Kalman filters are recursive algorithms that estimate the state of a dynamic system from a series of incomplete and noisy measurements. In the context of voice fidelity, the "state" could represent the clean voice signal, and the "measurements" could be the noisy received signal. The Kalman filter uses a mathematical model of the system to predict the next state and then updates this prediction based on the new measurement. This iterative process allows the filter to gradually refine its estimate of the true signal, even in the presence of significant noise. A key advantage of Kalman filters is their ability to handle non-stationary noise, meaning noise whose statistical properties change over time. However, they require a precise mathematical model of the system, which can be challenging to obtain in complex environments. Furthermore, Kalman filters can be computationally intensive, especially for high-dimensional systems. Despite these challenges, they can deliver outstanding performance in demanding applications, such as aerospace communications and medical signal processing. Picture trying to reconstruct a clear voice signal from the echoes and distortions of a long-distance phone call, the Kalman filter can leverage its understanding of the communication system to make accurate predictions about the signal, gradually refining the voice.

Moving on from complex mathematical models, we can explore simpler yet highly effective filtering techniques. Statistical filters leverage the statistical properties of the noise and the desired signal to separate them. For example, if the noise is known to have a different frequency distribution than the voice signal, a filter can be designed to attenuate the frequencies associated with the noise while preserving the frequencies of the voice. One common example is the Wiener filter, which minimizes the mean square error between the estimated signal and the desired signal based on their power spectral densities. Statistical filters are relatively easy to implement and can provide significant noise reduction with minimal computational overhead. However, their performance depends heavily on the accuracy of the statistical assumptions about the noise and the signal. If these assumptions are violated, the filter's performance can degrade significantly. To illustrate, think about removing the consistent hum of a poorly grounded electrical appliance from a recording; by analyzing the frequency spectrum and identifying the hum's characteristics, a statistical filter can effectively suppress it without altering the natural tone of the voice.

Finally, let's explore a crucial aspect of filter implementation: choosing the right parameters. No filter, however advanced, will perform optimally without careful tuning of its parameters. This includes selecting the appropriate filter order, cutoff frequencies, and windowing functions. The filter order determines the filter's steepness of the transition band, the region where the filter transitions from passing frequencies to attenuating them. A higher filter order provides a sharper transition but also increases the filter's complexity and computational cost. Cutoff frequencies define the boundaries between the passband and the stopband. Selecting these frequencies requires careful consideration of the frequency content of both the desired signal and the noise. Windowing functions are used in the design of Finite Impulse Response (FIR) filters to reduce unwanted artifacts, such as ringing and overshoot. Different windowing functions offer different trade-offs between main lobe width and side lobe attenuation. In practice, parameter selection often involves a process of experimentation and optimization. One approach is to use a training set of noisy voice signals and evaluate the filter's performance for different parameter settings. Another approach is to use an optimization algorithm to automatically search for the optimal parameter values.

Consider a scenario where you are adjusting the settings on an equalizer to emphasize a certain frequency, adjusting the parameters of a filter is very similar, it allows you to refine its performance to perfectly match your needs.

3.3 Adaptive Noise Cancellation

Adaptive Noise Cancellation (ANC) represents a sophisticated approach to mitigating unwanted sounds and enhancing voice fidelity. It's a dynamic process, far surpassing static filtering, as it continuously analyzes the incoming signal and adapts its noise reduction strategy in real-time. It's not a magic bullet, but rather a carefully constructed shield against the cacophony that often plagues voice transmissions. The beauty of ANC lies in its adaptability; unlike fixed filters that target specific frequency ranges, ANC intelligently identifies and neutralizes a wide range of noise sources, making it invaluable in dynamic environments.

The fundamental principle behind ANC involves generating an "anti-noise" signal that is precisely the inverse of the unwanted noise. This anti-noise, when combined with the original noisy signal, effectively cancels out the noise component, leaving behind a cleaner, more intelligible voice signal. To achieve this, ANC systems typically employ one or more microphones to capture the ambient noise. The signal from these microphones is then processed by an adaptive filter, which learns the characteristics of the noise and generates the corresponding anti-noise signal. The adaptive filter is the heart of the ANC system, constantly updating its parameters to match the evolving noise environment. It is a continuous dance between noise detection, analysis, and countermeasure creation.

Several factors influence the effectiveness of ANC. Microphone placement is paramount; the noise-sensing microphone must be positioned in a location where it accurately captures the ambient noise, but is minimally affected by the desired voice signal. If the microphone picks up too much of the voice, the ANC system may inadvertently cancel out portions of the voice signal, leading to distortion. The characteristics of the adaptive filter also play a crucial role. The filter's complexity, adaptation speed, and stability all impact its ability to accurately model and cancel the noise. A poorly designed filter may introduce artifacts or even amplify the noise in certain frequency ranges. Furthermore, the acoustic environment itself can influence ANC performance. Reverberation and reflections can create complex noise patterns that are difficult to cancel effectively. Despite these challenges, careful design and implementation can yield significant improvements in voice fidelity.

There are different implementations of ANC, each suited for specific applications. Feedforward ANC uses a reference microphone to capture the noise before it reaches the primary microphone. This approach is particularly effective in situations where the noise source is relatively predictable and the delay between the reference microphone and the primary microphone is minimal. Feedback ANC, on the other hand, uses the primary microphone to capture both the desired voice signal and the noise. The adaptive filter then estimates the noise component and generates the anti-noise signal. Feedback ANC is more challenging to implement due to the potential for instability, but it can be effective in situations where a reference microphone is not feasible. Hybrid ANC systems combine both feedforward and feedback techniques to achieve superior performance in a wider range of environments. The key is to select the appropriate architecture based on the specific application requirements and constraints.

Implementing ANC in real-world applications requires careful consideration of several practical aspects. The selection of microphones, amplifiers, and digital signal processors

(DSPs) is crucial for achieving optimal performance. The microphones must have low self-noise and high sensitivity to accurately capture the ambient noise. The amplifiers must be linear and have sufficient gain to amplify the microphone signals without introducing distortion. The DSP must have enough processing power to execute the adaptive filtering algorithms in real-time. Furthermore, the software implementation of the ANC algorithms must be carefully optimized for efficiency and stability. This often involves trade-offs between computational complexity and performance. For example, more complex adaptive filtering algorithms may provide better noise cancellation, but they also require more processing power.

Beyond the technical aspects, there are also ergonomic and aesthetic considerations. In applications such as headsets and hearing aids, the size and weight of the ANC system are important factors. The system must be comfortable to wear and unobtrusive. The placement of microphones and speakers must also be carefully considered to minimize feedback and maximize noise cancellation. Finally, the user interface should be intuitive and easy to use. Users should be able to easily adjust the level of noise cancellation to suit their individual preferences and the surrounding environment. The ultimate goal is to create an ANC system that seamlessly integrates into the user's workflow and enhances their communication experience.

To illustrate the power of ANC, consider its application in a noisy factory setting. A worker wearing a headset equipped with ANC can clearly communicate with colleagues or supervisors, despite the loud machinery and equipment. The ANC system effectively cancels out the background noise, allowing the worker to hear and be heard without straining their voice. This not only improves communication clarity but also reduces the risk of miscommunication and errors. Another example is in telemedicine, where clear audio transmission is critical for accurate diagnosis and treatment. ANC can eliminate ambient noise in both the patient's and the doctor's environments, ensuring that vital information is conveyed accurately. These are just a few examples of how ANC can make a significant difference in various real-world applications.

Finally, one must understand the limitations. While adaptive noise cancellation is undeniably a powerful tool, it's crucial to acknowledge its boundaries. It's not a perfect silencer; it can significantly reduce noise, but rarely eliminate it completely. Certain types of noise, particularly those with highly unpredictable or rapidly changing characteristics, can be challenging to cancel effectively. Moreover, ANC systems can sometimes introduce artifacts or distortions into the audio signal, especially if they are not properly designed or calibrated. It's essential to weigh the benefits of noise reduction against the potential for these side effects. Ultimately, ANC is just one tool in the arsenal for optimizing voice fidelity, and it should be used judiciously in conjunction with other techniques, such as proper acoustic design and careful microphone placement. By understanding both the strengths and limitations of ANC, we can harness its power to create clearer, more intelligible voice communication systems.

3.4 Impulse Noise Reduction

Impulse noise, often characterized by sudden, sharp bursts of interference, poses a unique challenge to voice fidelity. Unlike continuous background noise which can be addressed with conventional filtering, impulse noise manifests as transient spikes, disrupting the smoothness and clarity of the intended audio. Common sources include electrical switching, lightning strikes, and even physical impacts near the audio equipment. Its sporadic nature requires specialized reduction techniques to effectively mitigate its impact on the perceived audio quality. Simply increasing the signal strength often exacerbates the problem, amplifying both the desired signal and the unwanted impulses. This chapter will equip you with several methods to recognize and suppress this pervasive form of distortion, bringing clarity back to your audio signals.

One of the most effective methods to deal with impulse noise is median filtering. This non-linear filtering technique excels at removing impulse noise while preserving sharp edges and details in the signal, unlike linear filters which can blur these features. The core principle involves replacing each sample in the signal with the median value of its neighboring samples within a defined window. Let's say you have an audio sample series: [2, 5, 1, 8, 3, 9, 4]. If you apply a median filter with a window size of 3, you'd look at the first three values: [2, 5, 1]. The median of these is 2. So, the value '5' (the center of the window) would be replaced by '2'. Stepping along, the next window is [5, 1, 8], with a median of 5. This process is repeated for each sample in the data. Outliers, like those caused by impulse noise, are significantly different from their surrounding samples and are therefore effectively suppressed by the median calculation. The key is to select an appropriate window size. A larger window provides more aggressive noise reduction but can also blur finer details in the audio signal. Conversely, a smaller window preserves more detail but might not effectively remove wider impulses. Experimentation is crucial to finding the optimal balance for a specific application and audio characteristics. For instance, in speech, a relatively small window (e.g., 3-5 samples) may suffice, while in music, a slightly larger window could be used to address more pronounced impulse artifacts without excessive blurring.

Another robust approach to reducing impulse noise is through the application of thresholding techniques. This method involves identifying signal samples that exceed a predefined amplitude threshold, indicative of an impulse event. Once these outliers are detected, they can be either clipped to the threshold value or replaced with an estimated value based on the surrounding samples. There are several ways to determine the threshold. A simple approach is to use a fixed threshold based on the expected signal amplitude. However, a more adaptive method would be to calculate a dynamic threshold based on the local signal statistics, such as the moving average or standard deviation. This allows the threshold to adjust to variations in the background noise level, improving the accuracy of impulse detection. After detecting the impulse, several replacement strategies are possible. Clipping the signal to the threshold level is the simplest, but can still introduce audible artifacts. A more refined technique involves interpolating the impulse sample using the neighboring values. Linear interpolation, for example, calculates a new value based on the average of the samples immediately before and after the impulse. More sophisticated interpolation methods, such as spline interpolation, can provide smoother transitions and further reduce artifacts. It is important to consider the characteristics of the signal when choosing a replacement strategy. Clipping may be acceptable for certain types of signals, while interpolation is generally preferred for more sensitive applications like high-fidelity audio.

Moving beyond static techniques, adaptive filtering offers a dynamic approach to mitigating impulse noise. Adaptive filters, such as the Least Mean Squares (LMS) filter, learn the characteristics of the noise and automatically adjust their filtering parameters to minimize its impact. To implement adaptive impulse noise reduction, a reference noise signal is required. This signal should ideally be correlated with the impulse noise but uncorrelated with the desired voice signal. In practice, this might involve using a separate microphone positioned to capture the ambient noise environment, or by synthesizing a noise signal based on known characteristics of the impulse sources. The adaptive filter then uses the reference noise signal to estimate the impulse noise component in the primary audio signal. This estimate is subtracted from the primary signal, effectively canceling out the noise. A crucial element of adaptive filtering is the learning rate parameter. This parameter controls how quickly the filter adapts to changes in the noise characteristics. A higher learning rate allows the filter to track rapidly changing noise, but can also make it more susceptible to instability. Conversely, a lower learning rate provides greater stability, but may struggle to track fast-varying noise. Fine-tuning the learning rate is essential to achieving optimal performance. Adaptive filtering is particularly effective in scenarios where the characteristics of the impulse noise are time-varying or unknown. Its ability to learn and adapt to the noise environment makes it a powerful tool for achieving high-fidelity voice reception.

Frequency domain processing provides another powerful set of tools for impulse noise reduction. By transforming the audio signal into the frequency domain, it becomes possible to identify and suppress impulse noise based on its spectral characteristics. Typically, impulse noise manifests as a broad spectrum of frequencies, spread across a wide range. This contrasts with the narrower spectral content of typical voice signals. One technique involves using a spectral subtraction method. This approach estimates the noise spectrum during periods when the voice signal is absent. This noise estimate is then subtracted from the overall spectrum when the voice signal is present, effectively suppressing the noise components. A key challenge in spectral subtraction is accurately estimating the noise spectrum without removing important components of the voice signal. Overestimation of the noise spectrum can lead to audible artifacts, such as a "musical noise" effect. Adaptive techniques, such as Wiener filtering, can improve the accuracy of spectral subtraction by taking into account the signal-to-noise ratio at each frequency. Another approach is to use spectral gating. This method involves setting a threshold for each frequency bin. Any frequency components that fall below the threshold are suppressed, effectively removing low-amplitude noise. Spectral gating can be particularly effective for removing impulse noise that has a distinct spectral signature. Frequency domain processing offers a flexible and powerful approach to impulse noise reduction, enabling targeted suppression of noise components based on their spectral characteristics.

Finally, consider the impact of pre-processing techniques. Before applying any of the aforementioned noise reduction methods, consider applying pre-processing techniques to enhance the signal and make it more amenable to noise reduction. One simple yet effective technique is pre-emphasis. This involves boosting the high-frequency components of the audio signal. High frequencies are often more susceptible to the effects of impulse noise. Pre-emphasis helps to improve the signal-to-noise ratio in these critical frequency ranges. However, pre-emphasis can also amplify high-frequency noise, so it should be used judiciously in conjunction with other noise reduction techniques. Another pre-processing step could involve time-frequency analysis using wavelets. Wavelet transforms can decompose the audio signal into different frequency bands at different time scales, enabling targeted noise reduction in specific time-frequency regions. By identifying and suppressing impulse noise components in the wavelet domain, it is possible to achieve a more precise and less intrusive form of noise reduction. The choice of pre-processing techniques will depend on the specific characteristics of the audio signal and the noise environment. By carefully selecting and applying these techniques, it is possible to significantly enhance the effectiveness of subsequent noise reduction methods. Integrating these methods allows a holistic approach to noise elimination.

3.5 Filter Implementation Strategies

Implementing filters to improve voice fidelity isn't just about choosing the right type; it's about the practical strategies you employ to bring those filters to life. Consider this: you have a symphony of options, from the simplest analog circuits to the most sophisticated digital algorithms, each with its own set of trade-offs and considerations. Your challenge, as a voice engineer, is to orchestrate these options to achieve the clearest, most faithful representation of the human voice possible, as Parchment's Whispers will help you achieve.

Let's start with the analog realm, where simplicity meets elegance. Passive filters, constructed from resistors, capacitors, and inductors, offer a straightforward approach to shaping the frequency response of a signal. A basic RC low-pass filter, for instance, can effectively attenuate high-frequency noise that might be present in a received voice signal. The beauty of these filters lies in their simplicity: they require no external power and introduce minimal distortion. However, their performance is limited by the passive components used. The values of resistors, capacitors, and inductors can drift over time and temperature, affecting the filter's cutoff frequency and overall response. Moreover, passive filters can introduce insertion loss, which means they can reduce the overall signal level. To counteract this, you might need to add an amplifier stage, which then introduces its own complexities. Active filters, on the other hand, use operational amplifiers (op-amps) in conjunction with resistors and capacitors to achieve more complex filtering characteristics. With active filters, you can create high-pass, band-pass, and band-stop filters with greater precision and control than you could with passive circuits. Op-amps provide gain, which can compensate for insertion loss and improve the signal-to-noise ratio. They also allow for more sophisticated filter designs, such as Butterworth, Chebyshev, and Bessel filters, each with its own unique trade-offs between passband ripple, stopband attenuation, and phase response. Think carefully about what you're trying to achieve. If you need a steep roll-off to remove a specific interfering frequency, a Chebyshev filter might be a good choice, despite its ripple in the passband. If you need a flat passband and linear phase response, a Bessel filter might be more appropriate, even if it has a slower roll-off. Careful component selection is vital when designing and implementing analog filters. Use high-quality resistors and capacitors with tight tolerances to ensure the filter performs as expected. Consider using film capacitors, which have better stability and lower distortion than ceramic capacitors. For inductors, choose components with low DC resistance to minimize insertion loss. Don't forget to consider the op-amp itself. Select an op-amp with low noise, high gain-bandwidth product, and sufficient slew rate to handle the frequencies present in the voice signal.

Moving into the digital domain opens up a whole new world of possibilities. Digital filters, implemented using digital signal processing (DSP) techniques, offer unparalleled flexibility and precision. Unlike analog filters, digital filters are not affected by component tolerances or drift. They can be easily modified and reconfigured in software, making them ideal for adaptive noise cancellation and other advanced filtering applications. One of the most common types of digital filters is the Finite Impulse Response (FIR) filter. FIR filters are known for their linear phase response, which means they do not introduce phase distortion into the signal. This is especially important for voice signals, where phase distortion can significantly degrade intelligibility. FIR filters are implemented by convolving the input signal with a set of filter coefficients. The coefficients determine the filter's frequency response. Designing an FIR filter involves selecting appropriate coefficients to achieve the desired filtering characteristics. This can be done using a variety of techniques, such as the window method, the frequency sampling method, or the Parks-McClellan algorithm. Each method has its own advantages and disadvantages in terms of computational complexity, filter performance, and design flexibility. On the other hand, Infinite Impulse Response (IIR) filters offer steeper roll-off characteristics than FIR filters with the same number of coefficients. This can be advantageous when you need to sharply attenuate a specific interfering frequency. However, IIR filters can have non-linear phase responses, which can introduce phase distortion. IIR filters are implemented using feedback, which means that the output of the filter depends not only on the current input but also on previous outputs. This feedback can make IIR filters more efficient to implement than FIR filters, but it also makes them more susceptible to instability. When implementing digital filters, you need to consider the trade-offs between computational complexity, filter performance, and memory requirements. FIR filters generally require more memory and processing power than IIR filters, but they offer better phase linearity and stability. The choice between FIR and IIR filters depends on the specific application and the available resources.

Adaptive filters take filtering to the next level, adjusting their characteristics dynamically based on the characteristics of the input signal. This is particularly useful for noise cancellation, where the noise signal is constantly changing. An adaptive filter typically uses a feedback loop to minimize the error between the desired signal and the actual output. The filter adjusts its coefficients iteratively until the error is minimized. The Least Mean Squares (LMS) algorithm is one of the most common adaptive filtering algorithms. The LMS algorithm updates the filter coefficients based on the instantaneous error between the desired signal and the actual output. While easy to implement, the LMS algorithm converges slowly. The Normalized Least Mean Squares (NLMS) algorithm improves the convergence speed of the LMS algorithm by normalizing the step size. The Recursive Least Squares (RLS) algorithm offers even faster convergence than the NLMS algorithm, but it is also more computationally complex. When implementing adaptive filters, you need to carefully choose the step size and the filter order. The step size determines how quickly the filter adapts to changes in the input signal. A large step size can lead to faster convergence but also increase the risk of instability. A small step size can improve stability but slow down convergence. The filter order determines the complexity of the filter. A higher filter order can provide better noise cancellation but also increase the computational complexity. Regardless of the type of filter you choose, you must consider practical implementation details. For analog filters, this means choosing appropriate components, designing the circuit layout carefully to minimize noise and interference, and testing the filter's performance thoroughly. For digital filters, this means selecting an appropriate DSP platform, optimizing the filter code for speed and efficiency, and validating the filter's performance using real-world voice signals. In both cases, remember that filtering is just one piece of the puzzle. To achieve truly exceptional voice fidelity, you must consider every aspect of the signal chain, from the microphone to the speaker.

4 Chapter 4: Modulation and Demodulation

Modulation, in essence, is the art of encoding our precious voice signals onto a carrier wave, a high-frequency signal better suited for long-distance travel. Without it, transmitting voice over the airwaves or through cables would be akin to whispering across a crowded room – futile and unheard. Several modulation schemes exist, each with its own strengths and weaknesses, and choosing the right one is paramount to achieving optimal voice fidelity, a fundamental principle that every communications engineer should be acquainted with. Amplitude Modulation (AM), Frequency Modulation (FM), and various digital modulation techniques form the basis of all communications systems in the world today.

AM, or Amplitude Modulation, is arguably the simplest form. It works by varying the amplitude of the carrier wave in direct proportion to the amplitude of the voice signal. Imagine a dimmer switch controlling the brightness of a light; the voice signal is the hand on the dimmer, and the carrier wave is the light itself. As your voice gets louder, the carrier wave's amplitude increases; as it gets softer, the amplitude decreases.

AM is easy to implement, but it is very susceptible to noise. Any interference that affects the carrier wave's amplitude will be directly translated into noise in the received voice signal. Think of it like a scratch on a vinyl record – it's a simple system, but easily marred by imperfections. AM is commonly used in broadcast radio, particularly for long-range communication where its simplicity and tolerance to certain types of interference make it a viable choice despite its noise sensitivity.

Next we have FM, Frequency Modulation, that takes a different approach. Instead of altering the amplitude, FM varies the frequency of the carrier wave according to the amplitude of the voice signal. This makes FM inherently more immune to amplitude-based noise, as the demodulator only cares about changes in frequency, not changes in amplitude. Think of a violinist playing the same note but changing how fast the bow moves across the strings; that produces a wavering sound, that is what an FM is about. FM broadcasting is a popular choice in radio broadcasting since it provides high-fidelity and noise-free audio experiences. However, FM requires a wider bandwidth than AM, meaning it needs more space on the radio spectrum to transmit. This trade-off between noise immunity and bandwidth efficiency is a crucial consideration when choosing a modulation scheme.

Besides the analog modulation techniques already discussed, the world of digital modulation schemes offers a more robust set of features in terms of data transmission and immunity to external factors, although more complex than the previous ones. Phase-Shift Keying (PSK) and Quadrature Amplitude Modulation (QAM) are two prominent examples. PSK encodes information by varying the phase of the carrier wave, while QAM combines both phase and amplitude modulation to transmit even more data. These techniques are widely used in digital communication systems, like mobile phones and Wi-Fi, where high data rates and reliable transmission are paramount. Imagine morse code, but with hundreds of possible signals that change extremely quickly; that is somehow the concept of digital modulations.

After the signal has traveled through the air, and hopefully has reached the antenna, we need to get it back to what it was before, that is, we need to extract the voice from the modulated signal. Demodulation is the process of recovering the original voice signal from the modulated carrier wave. It is the inverse operation of modulation, and its success depends heavily on the type of modulation used and the quality of the received signal. For AM, demodulation is relatively straightforward; it involves simply detecting the peaks of the carrier wave, which correspond to the amplitude variations of the voice signal. However, for FM and digital modulation schemes, demodulation is significantly more complex, often requiring sophisticated circuits and algorithms.

A key challenge in demodulation is dealing with noise and interference that have been introduced during transmission. These unwanted signals can distort the carrier wave, making it difficult to accurately recover the original voice signal. Strategies for mitigating these challenges include using filters to remove out-of-band noise, employing synchronization techniques to track the carrier wave's frequency and phase, and implementing error correction codes to detect and correct errors in the received data. The ultimate goal is to minimize the impact of noise and interference on the demodulated voice signal, ensuring that it retains as much of the original fidelity as possible.

The art of optimizing demodulation is a constant pursuit, driven by the desire for clearer, more intelligible voice communication. One crucial aspect is synchronization. If the demodulator isn't perfectly synchronized with the carrier wave, distortion and data loss can occur. Advanced synchronization techniques, such as phase-locked loops (PLLs), are often employed to accurately track the carrier wave's frequency and phase, even in the presence of noise and interference. Error correction is another powerful tool for improving demodulation performance. By adding redundant information to the transmitted signal, error correction codes allow the demodulator to detect and correct errors that may have occurred during transmission. These codes are particularly effective in combating burst errors, where multiple consecutive bits are corrupted by noise.

Moreover, adaptive demodulation techniques can significantly improve performance in dynamic and challenging environments. These techniques automatically adjust the demodulator's parameters based on the characteristics of the received signal, such as its signal-to-noise ratio (SNR) and the type of interference present. For example, an adaptive demodulator might switch between different filtering strategies or error correction codes depending on the current conditions. This adaptability allows the demodulator to maintain optimal performance even in the face of varying noise levels and interference patterns.

Finally, the quality of the hardware used in the demodulation process plays a crucial role in its overall effectiveness. High-quality amplifiers, filters, and detectors can help to minimize noise and distortion, leading to a cleaner and more accurate demodulated voice signal. Furthermore, careful circuit design and layout can help to reduce the effects of electromagnetic interference (EMI), which can degrade demodulation performance. In summary, optimizing demodulation requires a holistic approach that considers synchronization, error correction, adaptive techniques, and high-quality hardware. By carefully addressing each of these aspects, it is possible to achieve significant improvements in voice fidelity and communication clarity.

4.1 Common Modulation Schemes

Modulation, at its core, is the art of impressing information onto a carrier wave. Think of it like sending a coded message; the carrier wave is the messenger, and the information we want to transmit – our voice – is the code. Several techniques achieve this, each with its strengths and weaknesses, particularly concerning voice fidelity. The right choice depends on the specific application, desired range, bandwidth limitations, and acceptable level of distortion.

Amplitude Modulation (AM), one of the simplest forms, alters the amplitude of the carrier wave according to the instantaneous amplitude of the voice signal. The louder you speak, the higher the amplitude of the transmitted signal, as simply as that. AM's simplicity comes at a price; it's highly susceptible to noise. Any variation in the amplitude of the signal, whether it's your voice or atmospheric interference, gets interpreted as part of the message. This is why AM radio can sound crackly, especially during storms. For voice fidelity, AM is generally the least preferred choice, except in situations where simplicity and cost are paramount and the receiving end is robust enough to discard disturbances.

Frequency Modulation (FM), a significant improvement over AM, encodes information by varying the frequency of the carrier wave. The amplitude remains constant. This makes FM inherently more resistant to amplitude noise. Static, crackling, and other unwanted amplitude variations are largely ignored by the FM receiver, leading to a cleaner, clearer audio signal. It's the reason FM radio sounds much better than AM. The trade-off, however, is that FM requires a wider bandwidth than AM. Each station needs more space in the radio spectrum, limiting the number of channels available. It is a worthwhile tradeoff, because it is better for high-fidelity voice transmission, and as a rule it is used in applications such as professional radio communications and high-quality audio broadcasting.

Phase Modulation (PM), closely related to FM, alters the phase of the carrier wave. Changes in the phase correspond to changes in the modulating signal. While PM is less common than FM for direct voice transmission, it serves as a crucial component in many digital modulation schemes. The difference between FM and PM might seem subtle, but it has to do with how the frequency or phase changes respond to the signal over time. They are mathematically linked, and in many digital systems, a combination of phase and amplitude modulation is used to maximize data throughput.

Moving into the realm of digital modulation, we encounter techniques specifically designed to transmit digital data, which, in the context of voice communication, often involves encoded voice signals. Digital modulation offers several advantages, including greater noise immunity, efficient use of bandwidth (with the right techniques), and the ability to incorporate error correction. The main digital modulation schemes you are likely to come across include Amplitude Shift Keying (ASK), Frequency Shift Keying (FSK), and Phase Shift Keying (PSK). Each represents digital information by changing, respectively, the amplitude, frequency, or phase of a carrier signal.

ASK is the simplest; it transmits data by varying the amplitude of the carrier wave. A high amplitude might represent a "1," while a low amplitude (or no signal at all) represents a "0." It's straightforward to implement but, like AM, it's susceptible to noise. FSK, on the other hand, uses different frequencies to represent different data bits. One frequency could be "1," and another could be "0." This is more robust than ASK as the frequency is less likely to get altered by noise compared to the amplitude.

PSK encodes information by varying the phase of the carrier wave. Binary Phase Shift Keying (BPSK) is the simplest form, using two phases (e.g., 0 degrees and 180 degrees) to represent "1" and "0." A more sophisticated variant, Quadrature Phase Shift Keying (QPSK), uses four phases, allowing it to transmit two bits of data per symbol, effectively doubling the data rate for the same bandwidth. There are even more complex forms of PSK with even more phases, providing even better data rate.

Quadrature Amplitude Modulation (QAM) takes things a step further by modulating both the amplitude and the phase of the carrier wave. By using multiple amplitude levels and multiple phase shifts, QAM can pack a significant amount of data into a single symbol. For instance, 16-QAM uses 16 different combinations of amplitude and phase, allowing it to transmit four bits per symbol. QAM is widely used in modern communication systems, including digital television, cable modems, and wireless networks, because it offers excellent bandwidth efficiency.

Selecting the right modulation scheme is a critical design decision. Consider the trade-offs. AM is simple but noisy. FM offers better audio quality but requires more bandwidth. Digital modulation techniques like QAM provide high data rates but require more complex hardware. In the context of "Parchment's Whispers", the goal is to preserve voice fidelity. This often means prioritizing noise immunity and distortion minimization. While advanced digital modulation schemes like

QAM are ideal for transmitting high data rates, they may not always be the best choice for voice, because voice codecs themselves must operate flawlessly to allow for perfect data transmission. FM, or robust digital modulation techniques with appropriate error correction, might be preferable for scenarios where voice clarity is paramount. Ultimately, testing and validation are essential to determine which modulation scheme delivers the best performance for a given application and the targeted audience.

Furthermore, remember that the choice of modulation scheme isn't isolated; it interacts with other components in the communication chain, such as the antenna, amplifier, and filters. A well-designed system optimizes all these elements to achieve the desired voice fidelity. For example, even the best modulation scheme will suffer if the antenna is poorly matched or the amplifier introduces excessive noise.

4.2 Demodulation Challenges

Demodulation, the process of extracting the original information signal from a modulated carrier wave, presents a unique set of obstacles to achieving high voice fidelity. This is where the rubber meets the road in signal recovery, and seemingly minor imperfections in the demodulation process can have a cascading effect on the perceived clarity and quality of the received voice. Unlike the relatively static challenges in signal transmission, demodulation often requires adapting to dynamically changing signal conditions and unpredictable interference.

One of the primary challenges arises from synchronization errors. Accurate demodulation hinges on precise timing synchronization between the receiver and the transmitter. In essence, the receiver needs to "know" exactly when each symbol or bit of information begins and ends. Any deviation from this perfect synchronization, no matter how small, can lead to bit errors and distortions in the recovered voice signal. Consider the case of Frequency Shift Keying (FSK) demodulation; if the receiver's clock drifts even slightly, it might misinterpret the transmitted frequencies, leading to errors. Think of it like trying to read a sentence where the spaces between words are inconsistent and unpredictable – the meaning becomes difficult to decipher. These errors are further aggravated by the Doppler effect, which is predominant in mobile communication, causing frequency shifts. Jitter, a rapid and unpredictable variation in the timing of a digital signal, also contributes significantly to synchronization problems.

Another significant hurdle lies in noise and interference management. Even with effective filtering techniques, some level of noise invariably finds its way into the received signal. This noise can mask the subtle variations in the modulated carrier wave, making it difficult for the demodulator to accurately discern the intended information. Compounding the problem is the presence of various forms of interference, ranging from electromagnetic interference (EMI) from nearby electronic devices to co-channel interference from other transmitters operating on the same or adjacent frequencies. Imagine trying to overhear a conversation in a crowded room with multiple people talking simultaneously; the voices blend together, making it difficult to focus on the specific conversation you're trying to follow. Noise, in its essence, represents the cacophony of that crowded room. Techniques like coherent demodulation, while effective in ideal conditions, are particularly susceptible to noise because they rely on accurately estimating the phase of the carrier wave, which can be easily corrupted by noise.

Channel impairments further complicate the demodulation process. Wireless communication channels are rarely ideal; they introduce a range of distortions and impairments that can significantly degrade signal quality. Multipath fading, caused by the signal arriving at the receiver via multiple paths with varying delays and amplitudes, is a common phenomenon. This can lead to signal cancellation and distortion, making it difficult for the demodulator to accurately recover the original information. Furthermore, the channel's frequency response may not be perfectly flat, meaning that different frequency components of the signal are attenuated differently. This can distort the signal's spectral characteristics, making demodulation more challenging. Consider the echo you might hear in a large, empty room; that's analogous to multipath fading, where delayed versions of the signal interfere with the original, creating a blurred and distorted sound. To add complexity, demodulators should use channel estimation techniques that can be complex and resource intensive.

Complexity and computational limitations can pose a significant challenge, particularly in resource-constrained devices. Sophisticated demodulation techniques, such as those employing advanced equalization algorithms or iterative decoding schemes, can offer superior performance in challenging environments. However, these techniques often require significant computational resources, making them impractical for implementation in low-power devices with limited processing capabilities. Striking a balance between demodulation performance and computational complexity is crucial, especially in applications where power consumption and cost are major considerations. Demodulation algorithms must be optimized to minimize computational load without sacrificing accuracy.

Adaptive demodulation is vital for overcoming these dynamic challenges. The key is to adapt the demodulation process to the prevailing signal conditions. This requires continuously monitoring the signal quality and adjusting the demodulation parameters accordingly. For example, if the receiver detects a sudden increase in noise levels, it might switch to a more robust demodulation scheme that is less susceptible to noise. Or, if the receiver detects significant multipath fading, it might activate an equalization algorithm to compensate for the channel distortions. Adaptive demodulation requires sophisticated signal processing techniques and real-time decision-making capabilities. Demodulation parameters such as the carrier frequency, timing offset, and equalization coefficients must be dynamically adjusted to optimize performance in the face of changing conditions.

Finally, implementation imperfections can derail the best-laid plans. Even with a theoretically perfect demodulation algorithm, real-world implementations are invariably subject to imperfections. Component tolerances, circuit noise, and quantization errors can all degrade the performance of the demodulator. Careful design and meticulous calibration are essential to minimize the impact of these imperfections. Analog-to-digital converters (ADCs) and digital-to-analog converters (DACs) can introduce non-linearities and quantization noise. Similarly, imperfections in the local oscillator can lead to frequency drift and phase noise. The selection of high-quality components and the implementation of appropriate calibration procedures are critical for mitigating these implementation-related challenges. Therefore, these problems must be analyzed with specialized debugging tools.

In conclusion, achieving optimal voice fidelity requires a comprehensive approach to demodulation that addresses a wide range of challenges. From synchronizing with the transmitter to mitigating noise and interference, from compensating for channel impairments to managing computational complexity, and from adapting to changing signal conditions to minimizing implementation imperfections, each aspect of the demodulation process must be carefully considered and optimized. By understanding these challenges and implementing appropriate solutions, it is possible to extract the clear and pristine voice signal hidden within the modulated carrier wave, fulfilling the promise of high-fidelity communication.

4.3 Optimizing Demodulation

Demodulation, at its core, is the intricate dance of reversing the effects of modulation, extracting the original information signal from the carrier wave. However, achieving pristine voice fidelity during this process demands a multifaceted approach, one that considers the specific modulation scheme, the noise environment, and the capabilities of the demodulation circuitry itself. A naive approach to demodulation often yields a signal riddled with artifacts, undermining the efforts expended in earlier stages of the signal chain.

Let's consider Amplitude Modulation (AM) as a starting point. While seemingly simple, AM demodulation can suffer from envelope distortion, especially when dealing with strong carrier signals or rapid changes in the modulating signal. Traditional diode detectors, while inexpensive and easy to implement, are susceptible to these distortions. A more sophisticated approach involves synchronous detection, where a locally generated carrier signal, phase-locked to the incoming carrier, is used to multiply the AM signal. This method significantly reduces distortion and improves signal-to-noise ratio, but requires a stable and accurate phase-locked loop (PLL). Remember that the quality of the PLL directly impacts the effectiveness of synchronous detection. Jitter in the PLL will translate to noise in the demodulated signal.

Frequency Modulation (FM) presents its own set of challenges. The most common FM demodulation technique, using a frequency discriminator like a Foster-Seeley discriminator or a ratio detector, relies on converting frequency variations into amplitude variations. However, these discriminators are sensitive to amplitude variations in the received signal, which can be caused by multipath fading or interference. To mitigate this, a limiter stage is typically employed before the discriminator to remove amplitude variations. A well-designed limiter is crucial; excessive limiting can introduce distortion, while insufficient limiting leaves the discriminator vulnerable to noise. A superior method involves using a PLL as an FM demodulator. The PLL locks onto the incoming FM signal, and the voltage required to maintain the lock directly corresponds to the frequency variations, thus demodulating the signal. PLL-based FM demodulators offer excellent linearity and noise performance, making them ideal for high-fidelity voice communication.

Digital modulation schemes, such as Phase-Shift Keying (PSK) and Quadrature Amplitude Modulation (QAM), require coherent demodulation. This means the receiver must have precise knowledge of the carrier frequency and phase. Carrier recovery techniques, such as Costas loops and squaring loops, are used to estimate the carrier phase. The accuracy of carrier recovery is paramount for successful demodulation. Errors in phase estimation lead to symbol errors, which degrade voice quality. Furthermore, timing recovery is also crucial in digital demodulation. Accurate timing synchronization ensures that the received symbols are sampled at the optimal points in time. Mismatched timing can cause inter-symbol interference (ISI), where one symbol bleeds into the next, further increasing the error rate. Adaptive equalization techniques can be employed to combat ISI, especially in channels with significant multipath propagation. Equalizers attempt to undo the distortion caused by the channel, improving the signal quality at the demodulator input. These adaptive equalizers constantly change to adapt the variations in the channel conditions, by estimating the parameters of the channel.

Beyond the specific demodulation technique, several general principles apply to optimizing demodulation for voice fidelity. Firstly, minimizing noise at the input of the demodulator is critical. This involves careful selection of low-noise amplifiers (LNAs) and filters in the front-end receiver circuitry, as already addressed in previous chapters. The demodulator cannot recover information buried beneath the noise floor. Secondly, the demodulator's bandwidth should be carefully matched to the bandwidth of the modulated signal. An overly wide bandwidth admits unnecessary noise, while an overly narrow bandwidth distorts the signal. Adaptive bandwidth control can be beneficial in scenarios where the signal bandwidth varies. Thirdly, the demodulator's linearity is crucial for preserving the fidelity of the voice signal. Non-linearities introduce harmonics and intermodulation distortion, degrading the perceived audio quality. Using components with high linearity ratings and careful circuit design can help minimize these distortions.

Calibration and alignment are also essential for optimal demodulation performance. Any offsets or imbalances in the demodulator circuitry can lead to signal distortion. Regularly calibrating the demodulator, using known test signals, can ensure accurate operation. Furthermore, careful alignment of the demodulator's various stages, such as the PLL or the frequency discriminator, is crucial for maximizing performance. This can involve adjusting component values or tweaking circuit parameters to optimize the demodulator's response.

In summary, optimizing demodulation for voice fidelity requires a holistic approach that considers the modulation scheme, the noise environment, and the demodulator's design and calibration. By carefully selecting the appropriate demodulation technique, minimizing noise, ensuring linearity, and performing regular calibration, we can extract the original voice signal with minimal distortion and achieve the desired clarity and fidelity. Don't neglect the importance of regular maintenance and troubleshooting to ensure the ongoing performance of your demodulation system. Remember that even the best-designed system can degrade over time due to component aging or environmental factors.

5 Chapter 5: Encoding for Clarity

Voice encoding is the art of transforming the complex tapestry of human speech into a format that can be efficiently transmitted, stored, and ultimately, faithfully reproduced. This is not merely about converting analog sound waves into digital bits; it's about strategically shaping those bits to optimize for clarity, bandwidth, and robustness against the inevitable imperfections of the communication channel. Choosing the right codec, understanding its nuances, and applying appropriate pre\item and post-processing techniques are crucial steps in achieving crystal-clear voice communication. The initial step in any voice encoding process is analog-to-digital conversion (ADC), a pivotal moment where the continuous sound wave is sampled and quantized. The sampling rate, measured in Hertz (Hz), determines how frequently the sound wave's amplitude is measured, and the Nyquist-Shannon sampling theorem dictates that the sampling rate must be at least twice the highest frequency present in the signal to avoid aliasing. For voice communication, a sampling rate of 8 kHz is often sufficient for capturing intelligible speech, while

16 kHz or higher is preferred for wideband audio that preserves more of the natural nuances and harmonics of the human voice. Quantization, on the other hand, determines the number of discrete levels used to represent the amplitude of each sample. A higher number of levels, typically expressed in bits (e.g., 16-bit or 24-bit), results in greater precision and lower quantization noise, but also requires more data.

Beyond the fundamental ADC process, the choice of voice codec plays a dominant role in shaping the final audio quality and bandwidth requirements. Codecs, short for "coder-decoders," are algorithms that compress and decompress audio data. They employ a variety of techniques, ranging from simple waveform coding to sophisticated source coding, to achieve different trade-offs between compression ratio, audio quality, and computational complexity. Waveform codecs, such as Pulse Code Modulation (PCM) and Adaptive Differential PCM (ADPCM), aim to faithfully reproduce the original waveform. PCM is the simplest form of waveform coding, directly encoding each sample without any compression. ADPCM, a more advanced technique, exploits the redundancy in speech signals by predicting the next sample based on previous samples and only encoding the difference between the predicted and actual values. These codecs generally offer good audio quality at relatively high bit rates, making them suitable for applications where bandwidth is not a primary constraint. Source codecs, such as Linear Predictive Coding (LPC) and Code Excited Linear Prediction (CELP), take a different approach by attempting to model the characteristics of the human vocal tract and the excitation source (e.g., vocal cords or turbulent air). LPC analyzes the speech signal to estimate the parameters of a linear filter that represents the vocal tract, while CELP uses a codebook of excitation signals to synthesize the speech. These codecs can achieve significantly higher compression ratios than waveform codecs, but may introduce some artifacts or distortions, particularly at very low bit rates. Modern voice communication systems often employ hybrid codecs that combine elements of both waveform and source coding to achieve a balance between audio quality and bandwidth efficiency. Examples of such codecs include G.729, AMR (Adaptive Multi-Rate), and Opus. Error correction techniques are essential for ensuring reliable voice communication, particularly in noisy or unreliable channels. Even with the best encoding strategies, data loss or corruption can occur during transmission, leading to audible artifacts or even complete loss of speech intelligibility. Forward Error Correction (FEC) codes add redundant information to the encoded data, allowing the receiver to detect and correct errors without requiring retransmission. Common FEC codes used in voice communication include Reed-Solomon codes and convolutional codes. Automatic Repeat Request (ARQ) protocols, on the other hand, rely on retransmission to recover from errors. The receiver detects errors and requests the sender to retransmit the corrupted data packets. ARQ protocols can provide very high reliability but may introduce delay, making them less suitable for real-time voice communication. A hybrid approach that combines FEC and ARQ can provide a good balance between reliability and delay. For instance, FEC can be used to correct minor errors, while ARQ is reserved for more severe cases where FEC alone is insufficient.

In addition to choosing the right codec and employing error correction techniques, pre\item and post-processing steps can significantly enhance voice clarity. Pre-processing techniques applied at the encoding stage can improve the quality of the input signal before it is compressed. Noise reduction algorithms can remove background noise and interference, improving the signal-to-noise ratio (SNR) and making the speech signal easier to encode. Voice activity detection (VAD) algorithms can identify and remove periods of silence, reducing the amount of data that needs to be encoded. Automatic gain control (AGC) algorithms can adjust the signal level to optimize the dynamic range of the ADC. Post-processing techniques applied at the decoding stage can further enhance the audio quality. Speech enhancement algorithms can reduce residual noise and artifacts introduced by the codec. Packet loss concealment (PLC) algorithms can reconstruct missing or corrupted audio packets, minimizing the impact of data loss on the perceived audio quality. Dithering, a technique that adds a small amount of random noise to the signal, can reduce quantization noise and improve the perceived smoothness of the audio.

Selecting the right encoding strategy for voice communication involves a careful consideration of several factors, including bandwidth constraints, audio quality requirements, computational resources, and channel characteristics. The choice of codec should be based on a thorough evaluation of its performance in these different dimensions. Subjective listening tests can be used to assess the perceived audio quality of different codecs under various conditions. Objective metrics, such as Perceptual Evaluation of Speech Quality (PESQ) and Mean Opinion Score (MOS), can provide a quantitative measure of audio quality. Bit rate requirements should be carefully considered to ensure that the encoded voice signal can be transmitted within the available bandwidth. Computational complexity should be assessed to ensure that the codec can be implemented efficiently on the target hardware platform. Robustness to errors and packet loss should be evaluated to ensure reliable voice communication in noisy or unreliable channels. By carefully considering these factors and applying appropriate encoding and processing techniques, it is possible to achieve crystal-clear voice communication even under challenging conditions, allowing the nuances of "Parchment's Whispers" to be heard with pristine clarity.

5.1 Voice Encoding Basics

Voice encoding stands as a cornerstone in the pursuit of optimal voice fidelity, acting as the bridge between the analog world of human speech and the digital realm of transmission and storage. Understanding its nuances is crucial for anyone seeking to deliver pristine voice communication. The objective of voice encoding is deceptively simple: to convert the continuous, fluctuating waveform of sound into a discrete, digital representation that can be transmitted efficiently and reconstructed faithfully at the receiving end. However, the complexities arise when trying to achieve this without introducing distortions, losing information, or consuming excessive bandwidth. The chapter introduces the basic concepts of voice encoding and aims to provide a solid foundation for understanding the principles and techniques involved in converting voice signals into digital data.

The first step in voice encoding is sampling, which involves taking snapshots of the analog voice signal at regular intervals. The Nyquist-Shannon sampling theorem dictates that the sampling rate must be at least twice the highest frequency component present in the voice signal to accurately reconstruct it. For human speech, which typically ranges from 300 Hz to 3.4 kHz, a sampling rate of 8 kHz is commonly used, as it captures the essential frequencies for intelligibility. However, higher sampling rates, such as 16 kHz or 44.1 kHz (used in CD-quality audio), offer improved fidelity and can capture a wider range of frequencies, resulting in a more natural-sounding reproduction. The choice of sampling rate involves a trade-off between fidelity and bandwidth, as higher sampling rates require more data to be transmitted and stored. After the signal is sampled, quantization occurs. Quantization assigns a discrete value to each sample based on its amplitude. The number of quantization levels determines the precision with which the signal amplitude is represented. A higher number of levels results in finer granularity and reduced quantization error, which manifests as noise in the reconstructed signal. The bit depth, which is the number of bits used to represent each sample, directly influences the number of quantization levels. For example, 8-bit quantization provides 256 levels, while 16-bit quantization offers 65,536 levels, resulting in a significant improvement in signal-to-noise ratio and dynamic range. Quantization introduces an unavoidable error that is reduced by increasing the bit depth; this is called "quantization noise" and it must be considered during the process.

Following the first two steps, the encoding process itself is crucial. Once the voice signal has been sampled and quantized, it's time to encode it, aiming to reduce the amount of data needed for transmission or storage. This is where voice codecs, short for coder-decoders, come into play. Codecs employ various algorithms to compress the digital voice signal, removing redundancies and irrelevant information. Pulse Code Modulation (PCM) is the simplest form of voice encoding, directly representing each sample with a fixed number of bits, but it's not very efficient in terms of bandwidth usage. Differential Pulse Code Modulation (DPCM) improves upon PCM by encoding the difference between successive samples, rather than the absolute value of each sample. This takes advantage of the fact that voice signals tend to change relatively slowly over time, allowing for a reduction in the number of bits required to represent each difference value. More advanced codecs, such as Adaptive Differential Pulse Code Modulation (ADPCM), further enhance compression by dynamically adjusting the quantization step size based on the characteristics of the voice signal. This allows for a better trade-off between compression and quality, adapting to the varying dynamics of speech. Vector Quantization, transform coding, and sub-band coding are also popular.

However, encoding is not the last step. Error correction is a vital consideration in voice encoding, especially when transmitting over noisy or unreliable channels. Errors can occur due to various factors, such as interference, signal fading, or network congestion, leading to distortions or dropouts in the reconstructed voice signal. To mitigate these effects, error correction techniques are employed to detect and correct errors introduced during transmission. Forward Error Correction (FEC) adds redundant information to the encoded voice signal, allowing the receiver to detect and correct errors without requiring retransmission. Cyclic Redundancy Check (CRC) is a common FEC technique that appends a checksum to each data packet, enabling the receiver to verify the integrity of the received data. In cases where errors cannot be corrected, error concealment techniques can be used to minimize the impact of the errors on the perceived voice quality. This may involve interpolating missing samples or substituting them with plausible values based on the surrounding context.

The choice of voice encoding technique depends on a variety of factors, including the desired level of fidelity, the available bandwidth, the computational complexity, and the robustness requirements. Simple codecs like PCM are easy to implement and require minimal processing power, but they consume significant bandwidth. More advanced codecs like ADPCM offer better compression but demand more computational resources. When selecting a voice codec, it's essential to consider the specific application and its constraints. For example, mobile communication systems often prioritize bandwidth efficiency to accommodate a large number of users, while high-definition audio conferencing systems prioritize fidelity to deliver a natural and immersive experience. In the end, by considering the factors listed in the chapter, anyone can easily create a voice encoding system that has a great signal reception for voice fidelity clarity.

5.2 Advanced Voice Codecs

Voice codecs stand as the sentinels guarding the clarity of our transmitted words. The journey from analog sound waves to digital streams and back again is fraught with peril; codecs are the navigators, charting a course that minimizes distortion and maximizes intelligibility. But not all codecs are created equal. While basic codecs prioritize bandwidth efficiency, advanced codecs strive for a delicate balance: compressing data for transmission without sacrificing the nuances of the human voice. Understanding the subtleties of these advanced techniques is crucial for anyone seeking to optimize voice communication systems.

The world of advanced voice codecs extends far beyond the simple pulse-code modulation (PCM) that forms the foundation of many digital audio systems. While PCM directly converts analog waveforms into digital samples, more sophisticated codecs employ a range of techniques to reduce the amount of data needed to represent the same audio information. Adaptive Differential Pulse Code Modulation (ADPCM), for example, focuses on encoding the difference between successive samples, rather than the absolute value of each sample. This is advantageous because the changes in speech signals are often smaller than the signals themselves, resulting in a more compact representation. But even ADPCM has its limitations, particularly when dealing with complex and rapidly changing audio. Therefore, it is generally seen as a legacy technology.

Code-Excited Linear Prediction (CELP) codecs represent a significant leap forward in voice compression technology. These codecs leverage a model of human speech production to predict the upcoming sound. The core idea is to decompose speech into two primary components: a slowly varying vocal tract filter and a rapidly changing excitation signal. The vocal tract filter, which models the shape of the mouth and throat, determines the resonant frequencies (formants) of the voice. The excitation signal, on the other hand, represents the energy source that drives the vocal tract, such as the vibration of the vocal cords. By separately encoding these two components, CELP codecs can achieve high compression ratios without sacrificing speech quality. Different CELP variants exist, like Algebraic CELP (ACELP) which further optimizes the encoding process by using algebraic codebooks, reducing the computational complexity and memory requirements, thereby improving overall coding efficiency. ACELP is ubiquitous in modern mobile communication standards, such as GSM, 3G, and LTE, because its algorithm strikes a balance between voice quality and computational efficiency making it suitable for real-time communication on resource-constrained devices.

Moving beyond CELP, we encounter transform codecs. These codecs operate in the frequency domain, transforming the time-domain speech signal into its spectral representation using techniques like the Discrete Cosine Transform (DCT). By analyzing the frequency components of the speech, transform codecs can identify and discard less important spectral elements, further reducing the data rate. This approach is particularly effective for compressing speech with a wide dynamic range and complex spectral content. The Modified Discrete Cosine Transform (MDCT) is another method, useful in audio compression due to its reduction of artifacts.

One key parameter affecting the quality of a voice codec is its bit rate. Higher bit rates generally translate to better audio quality, as more data is available to represent the nuances of the speech signal. However, higher bit rates also require more bandwidth for transmission. The choice of bit rate, then, becomes a trade-off between quality and bandwidth efficiency. Variable bit rate (VBR) codecs address this trade-off by dynamically adjusting the bit rate based on the complexity of the audio signal. When speech is relatively simple, such as during pauses or quiet passages, the codec reduces the bit rate to conserve bandwidth. Conversely, when speech is complex and dynamic, the codec increases the bit rate to maintain audio quality. Adaptive Multi-Rate (AMR) codecs, for example, offer a range of bit rates to choose from, allowing the system to dynamically adapt to changing network conditions and user preferences. AMR is frequently used in mobile communication systems.

Beyond the core encoding algorithms, advanced voice codecs often incorporate sophisticated pre\item and post-processing techniques to further enhance audio quality. Noise suppression algorithms, for example, can reduce background noise, while echo cancellation algorithms can eliminate echoes in two-way communication. These techniques work by analyzing the audio signal and identifying characteristics indicative of noise or echoes. Once identified, the algorithms can selectively attenuate or remove these unwanted components, resulting in a cleaner and more intelligible voice signal. Additionally, bandwidth extension techniques can artificially extend the frequency range of the encoded speech, improving its perceived quality, especially at low bit rates. Such a technique is Spectral Band Replication (SBR), which reconstructs the higher frequencies of a signal from the lower ones.

Selecting the appropriate voice codec for a given application depends on a number of factors, including the desired audio quality, available bandwidth, computational resources, and target platform. For applications where voice quality is paramount, such as high-end conferencing systems, codecs with higher bit rates and sophisticated algorithms may be preferred. For applications where bandwidth is limited, such as mobile communication systems, codecs with lower bit rates and simpler algorithms may be more suitable. The choice also depends on the capabilities of the devices used for encoding and decoding. Some codecs require significant computational power, making them unsuitable for resource-constrained devices like smartphones or embedded systems. Careful consideration of these factors is essential for choosing a codec that strikes the right balance between quality, efficiency, and practicality.

Understanding these techniques is essential for crafting communication systems where voices are not just heard, but truly understood.

5.3 Error Correction Techniques

Error correction is paramount in ensuring voice fidelity. After all, even the most pristine signal can be corrupted during transmission or storage. Imagine whispering a secret across a crowded room – the chances of it being heard perfectly are slim. Similarly, in voice communication systems, noise, interference, and hardware limitations can introduce errors into the encoded voice signal. The task is to minimize the impact of these errors, ideally to the point where they become imperceptible to the listener. Error correction isn't about magically removing noise; instead, it's about adding redundancy to the signal in a way that allows the receiver to detect and potentially correct errors that may occur.

Consider a simple scenario: you're sending a packet of voice data across a network. This packet could be anything from a snippet of a phone call to a command sent to a voice assistant. Along its journey, a few bits get flipped due to electromagnetic interference or a glitch in a router. Without error correction, the receiver would interpret the corrupted data, leading to distortion or even a complete loss of information. However, by strategically adding extra bits to the packet, we can create a system where the receiver can detect these errors and, in some cases, reconstruct the original data. This approach is like adding checkpoints along a route to ensure the final destination is reached even if detours or unexpected delays occur.

Several error correction techniques exist, each with its own strengths and weaknesses. One of the most fundamental is parity checking. In its simplest form, a parity bit is added to a block of data to indicate whether the number of '1' bits is even or odd. If the receiver detects a different parity than expected, it knows an error has occurred. While parity checking is easy to implement, it can only detect an odd number of errors within a block. This limitation makes it unsuitable for applications where errors are more likely to occur in bursts.

A step up from parity checking is Cyclic Redundancy Check (CRC). CRC involves dividing the data by a specific polynomial and appending the remainder (the CRC checksum) to the data. The receiver performs the same division and compares its calculated checksum with the received one. If they don't match, an error is detected. CRCs are much more powerful than parity checks, capable of detecting a wide range of error patterns, including burst errors, while maintaining a relatively low overhead. Think of CRC as a sophisticated form of verifying the integrity of a document by comparing its digital "fingerprint" against a known good version.

Moving beyond simple error detection, we encounter Forward Error Correction (FEC) techniques. These methods not only detect errors but also allow the receiver to correct them without requesting retransmission of the data. A common example of FEC is Reed-Solomon coding. Reed-Solomon codes are particularly effective at correcting burst errors, which are common in wireless communication channels where fading or interference can cause multiple consecutive bits to be corrupted. The price for this correction capability is increased overhead, as more redundant data needs to be transmitted. Another type of FEC is convolutional coding, which works by encoding the data stream based on a sliding window of previous inputs. This introduces redundancy in a different way than block codes like Reed-Solomon, and the decoding process is typically performed using the Viterbi algorithm, which finds the most likely sequence of transmitted bits given the received signal. Convolutional codes are well-suited for channels with additive white Gaussian noise (AWGN), a common model for background noise in communication systems. Imagine FEC as sending a message with built-in repair instructions, ensuring that even if parts of the message are damaged, the receiver can still piece together the original content.

The choice of error correction technique depends on several factors, including the characteristics of the communication channel, the acceptable level of overhead, and the computational complexity of the encoding and decoding algorithms. For example, in a low-bandwidth, high-noise environment, a robust FEC scheme like Reed-Solomon might be necessary, even at the cost of increased overhead. Conversely, in a relatively clean channel with abundant bandwidth, a simple error detection scheme like CRC might suffice. Consider also the power constraints of the device. Complex FEC schemes require more processing power, which can drain battery life in mobile devices. Therefore, there is a tradeoff between error correction performance and energy efficiency that needs to be carefully considered.

Implementing error correction involves several practical considerations. First, you need to choose the appropriate coding scheme based on your channel characteristics and performance requirements. Then, you need to integrate the encoder and decoder into your voice communication system. This may involve using dedicated hardware or software libraries for encoding and decoding. It is essential to test the performance of your error correction scheme under realistic channel conditions. This can be done through simulations or by conducting field trials. The goal is to determine the error rate and voice quality under various conditions and to fine-tune the error correction parameters to achieve the desired level of performance. For example, in a VoIP (Voice over Internet Protocol) application, you might need to adjust the FEC parameters based on the measured packet loss rate of the network. In conclusion, carefully considering and implementing appropriate error correction techniques is essential for achieving high voice fidelity in modern communication systems.

6 Chapter 6: Digital Signal Processing

Let's embark on the journey into Digital Signal Processing (DSP), a critical domain for enhancing voice fidelity. DSP involves using mathematical algorithms to manipulate digital signals, in our case, audio signals, to improve their quality or extract useful information. For the beginner, this may seem daunting, but understanding a few core concepts will unlock a world of possibilities for improving voice clarity. The fundamental idea is to convert analog audio signals into digital representations, process these representations using specialized algorithms, and then convert them back into analog audio. This conversion is achieved through Analog-to-Digital Converters (ADCs) and Digital-to-Analog Converters (DACs). The quality of these converters plays a vital role, with higher sampling rates and bit depths generally resulting in better fidelity, capturing more nuances in the sound. Understanding these basics is the foundation upon which we can build our knowledge of more advanced DSP techniques specifically aimed at voice enhancement.

Echo cancellation represents a cornerstone of voice communication systems, particularly crucial in scenarios involving hands-free devices or speakerphones where the speaker's voice is picked up by the microphone after being played through the loudspeaker, creating a loop. This echo not only distracts the speaker but also degrades the overall quality of the conversation for the other party. The challenge lies in distinguishing the legitimate incoming signal from the delayed and distorted version of the outgoing signal. Adaptive filters are commonly employed for this task. These filters dynamically estimate the echo path, the acoustic characteristics of the environment that shape the echo, and then subtract this estimated echo from the microphone signal. The filter coefficients, which define the filter's characteristics, are continuously adjusted based on the error signal, the difference between the actual microphone signal and the estimated echo. Algorithms like the Normalized Least Mean Squares (NLMS) are popular choices for updating these coefficients, balancing convergence speed, how quickly the filter adapts, and stability, preventing the filter from diverging and introducing artifacts. Effectively implementing echo cancellation requires careful consideration of factors such as echo path delay, reverberation time, and the presence of non-linear distortions, which can complicate the echo estimation process.

Venturing into speech enhancement techniques opens up an array of possibilities for cleaning up noisy voice signals. Noise reduction algorithms aim to suppress background noise while preserving the integrity of the desired speech signal. Spectral subtraction, a classic technique, estimates the noise spectrum during periods of silence and subtracts this estimate from the noisy speech spectrum. While effective in reducing stationary noise, spectral subtraction can introduce musical noise, artifacts that sound like short, tonal bursts, particularly when the noise estimate is inaccurate. Wiener filtering, based on statistical signal processing, offers a more sophisticated approach by estimating the speech and noise power spectral densities and designing a filter that minimizes the mean-squared error between the estimated speech and the clean speech. This technique generally produces fewer artifacts than spectral subtraction but requires accurate estimation of the signal and noise statistics. More advanced techniques, such as those based on deep learning, are emerging as powerful tools for speech enhancement, capable of handling complex noise environments and non-stationary noise sources with remarkable accuracy. The implementation of these techniques often involves trade-offs between noise reduction performance, speech distortion, and computational complexity.

Diving into spectral subtraction requires a closer look at its implementation. At its core, it relies on the assumption that noise is additive and can be estimated during pauses in speech. The algorithm first estimates the noise power spectrum during these silent intervals. Then, it subtracts this noise estimate from the power spectrum of the noisy speech signal. The challenge is to accurately estimate the noise spectrum and to avoid over-subtracting, which can lead to speech distortion. Different variations of spectral subtraction exist, each with its own strengths and weaknesses. For example, some implementations use different subtraction factors for different frequency bands, depending on the signal-to-noise ratio in that band. Others employ smoothing techniques to reduce the musical noise artifacts. Careful parameter tuning is essential to achieve a good balance between noise reduction and speech quality.

Moving on to Wiener filtering, its effectiveness hinges on the accuracy of the speech and noise power spectral density estimation. These estimates can be obtained using various methods, such as recursive averaging or Kalman filtering. The Wiener filter is then designed based on these estimates to minimize the mean-squared error between the estimated speech and the clean speech. This filter effectively attenuates frequencies where the noise power is high and preserves frequencies where the speech power is dominant. The Wiener filter offers improved performance compared to spectral subtraction, particularly in non-stationary noise environments. However, it is computationally more complex and requires careful selection of the estimation parameters to avoid introducing artifacts.

Exploring adaptive noise cancellation offers another powerful tool. These filters continuously adapt to the changing noise environment, making them particularly effective in non-stationary noise conditions. The core idea is to use a reference microphone to capture the noise signal and then subtract a filtered version of this noise signal from the primary microphone signal, which contains both speech and noise. The filter coefficients are adaptively adjusted using algorithms like the Least Mean Squares (LMS) or Recursive Least Squares (RLS) to minimize the error signal, the difference between the primary microphone signal and the estimated speech signal. Adaptive noise cancellation requires careful placement of the reference microphone to ensure that it captures a representative sample of the noise environment but does not pick up the desired speech signal.

To conclude, DSP provides a diverse toolkit for enhancing voice fidelity. From the fundamental principles of analog-to-digital conversion to the advanced algorithms for echo cancellation and noise reduction, mastering these techniques empowers engineers to create communication systems with unparalleled clarity and intelligibility. Experimenting with different algorithms, tuning parameters, and evaluating performance in real-world scenarios is key to unlocking the full potential of DSP for voice enhancement. Remember, the optimal approach often depends on the specific application, the characteristics of the noise environment, and the desired trade-offs between performance, complexity, and computational cost. Continuously exploring new techniques and refining existing ones will ensure that voice communication remains clear and understandable, regardless of the environment. The journey into DSP is ongoing, with new algorithms and techniques constantly emerging, driven by the ever-increasing demand for better voice communication.

6.1 DSP Fundamentals

Digital Signal Processing, or DSP, forms the backbone of modern voice communication systems. It's the art and science of manipulating signals – in our case, voice – using computational algorithms. Unlike analog signal processing, which relies on physical components like resistors and capacitors, DSP performs these operations using mathematical equations on digitized signals. This approach offers unparalleled flexibility, precision, and the ability to implement complex algorithms that are simply impossible in the analog domain. This is where the magic happens, and our voice signals are sculpted and refined.

The very first step in any DSP system is Analog-to-Digital Conversion (ADC). This crucial process transforms the continuous analog voice signal, captured by a microphone, into a discrete digital representation that a computer can understand. The ADC samples the analog signal at regular intervals, converting each sample into a numerical value. Two key parameters govern the quality of this conversion: sampling rate and bit depth. The sampling rate, measured in Hertz (Hz), determines how many samples are taken per second. The Nyquist-Shannon sampling theorem dictates that the sampling rate must be at least twice the highest frequency present in the signal to avoid aliasing – a phenomenon where high-frequency components are misinterpreted as lower frequencies, leading to distortion. For voice communication, a sampling rate of 8 kHz is commonly used, as it captures most of the speech spectrum. Higher sampling rates, like 16 kHz or 44.1 kHz (used in CDs), offer improved fidelity but require more processing power and storage space. Bit depth, on the other hand, determines the precision with which each sample is represented. A higher bit depth provides more quantization levels, resulting in a more accurate representation of the analog signal and reduced quantization noise. Common bit depths for voice communication include 8 bits and 16 bits. Once the analog signal is digitized, it enters the digital domain, ready for processing.

One of the fundamental operations in DSP is the Discrete Fourier Transform (DFT). The DFT decomposes a time-domain signal into its constituent frequencies, revealing the spectral content of the voice signal. This information is invaluable for tasks like noise reduction, speech recognition, and audio analysis. The DFT transforms a sequence of N complex numbers into another sequence of N complex numbers, representing the amplitudes and phases of the different frequencies present in the signal. While the DFT provides a powerful tool for spectral analysis, its computational complexity can be a bottleneck for real-time applications. The Fast Fourier Transform (FFT) is an efficient algorithm for computing the DFT, significantly reducing the computational burden. The FFT achieves this speedup by exploiting symmetries in the DFT equation, allowing for a much faster calculation, especially for large values of N. Understanding the DFT and FFT is essential for anyone working with DSP, as they form the basis for many spectral analysis and processing techniques. In practice, specialized hardware accelerators are often used to further accelerate FFT calculations, enabling real-time processing of high-bandwidth signals.

Beyond spectral analysis, DSP enables a wide range of signal processing techniques that are crucial for enhancing voice fidelity. Finite Impulse Response (FIR) filters and Infinite Impulse Response (IIR) filters are fundamental building blocks for shaping the frequency response of a signal. FIR filters are characterized by their finite impulse response, meaning that their output settles to zero after a finite amount of time in response to an impulse input. They are inherently stable and can be designed to have linear phase response, which preserves the shape of the signal in the time domain. IIR filters, on the other hand, have an infinite impulse response, meaning that their output continues indefinitely in response to an impulse input. They can achieve sharper frequency responses with fewer coefficients than FIR filters, but they are more susceptible to instability and non-linear phase distortion. Choosing the right type of filter depends on the specific application and the trade-offs between performance, stability, and computational complexity. In voice communication, filters are used to remove unwanted noise, equalize the frequency response of the channel, and shape the speech signal for optimal intelligibility.

Another essential DSP technique is adaptive filtering. Unlike fixed filters, adaptive filters can adjust their parameters based on the characteristics of the input signal. This is particularly useful in situations where the noise or interference is time-varying or unknown. Adaptive filters are commonly used for noise cancellation, echo cancellation, and channel equalization. The Least Mean Squares (LMS) algorithm is a popular adaptive filtering algorithm that iteratively adjusts the filter coefficients to minimize the mean squared error between the desired signal and the filter output. Other adaptive filtering algorithms include Recursive Least Squares (RLS) and Kalman filtering, which offer faster convergence and improved performance in certain scenarios but require more computational resources. Adaptive filtering is a powerful tool for enhancing voice fidelity in challenging environments, allowing the system to adapt to changing conditions and maintain optimal performance. We also rely on compression techniques to reduce the amount of data needed to transmit the audio signal. By removing redundant information we achieve efficient use of bandwidth, while keeping the signal quality high.

The journey from analog voice to clear digital communication is a sophisticated process. Understanding these DSP fundamentals will empower you to design and implement voice communication systems with unparalleled fidelity. This is not just about technology; it's about connecting people and sharing ideas with utmost clarity. By mastering these tools, you contribute to a world where every voice is heard. And what could be more rewarding?

6.2 Echo Cancellation Algorithms

Echoes, unwanted reflections of your own voice, can severely degrade the quality of a conversation, making communication frustrating and unnatural. Imagine trying to have a serious discussion while hearing yourself, delayed and distorted – not ideal, right? The algorithms designed to combat these echoes, therefore, form a cornerstone of modern communication systems, ensuring clear and natural sound quality.

The fundamental principle behind echo cancellation is prediction and subtraction. We essentially try to predict the echo that will be returned and then subtract that predicted echo from the received signal. Let's break down how this works. The system starts by analyzing the outgoing signal. Think of it as listening to what you're about to say. It then uses this information to build a model of the echo path – the route the sound takes from your speaker to the microphone, bouncing off surfaces along the way. This echo path includes all sorts of factors, like the distance between the speaker and microphone, the acoustic properties of the room, and even the characteristics of the microphone and speaker themselves. The system estimates how the outgoing signal will be altered as it travels through this echo path, effectively predicting what the echo will sound like when it comes back. Finally, the predicted echo is subtracted from the received signal, which contains both the far-end speech and the echo. If the prediction is accurate, the echo will be effectively canceled out, leaving only the desired signal. It sounds simple in theory, but the real challenge lies in accurately modeling that complex echo path.

One of the most common and effective echo cancellation algorithms is the Normalized Least Mean Squares (NLMS) algorithm. NLMS is an adaptive filter, meaning it continuously adjusts its parameters to improve its echo prediction. It starts with an initial guess of the echo path and then iteratively refines this estimate based on the error between the predicted echo and the actual echo present in the received signal. The "normalized" part of NLMS refers to a normalization factor that helps to stabilize the algorithm and prevent it from diverging, particularly when the input signal has a wide dynamic range. Imagine you're trying to focus a camera lens on a moving object. NLMS is like constantly tweaking the focus to keep the image sharp, even as the object moves and the lighting changes. The algorithm uses a "step size" parameter to control how quickly it adapts. A larger step size allows for faster adaptation, but it can also make the algorithm more susceptible to noise and instability. A smaller step size provides more stability but can be slower to adapt to changes in the echo path. Choosing the right step size is a crucial balancing act. NLMS is relatively simple to implement and computationally efficient, making it a popular choice for many applications, especially where processing power is limited.

However, NLMS isn't perfect. It struggles with double-talk scenarios, where both parties are speaking simultaneously. During double-talk, the algorithm can mistake the far-end speech for an echo and try to cancel it out, leading to signal distortion. To address this, echo cancellation systems often incorporate double-talk detectors. These detectors analyze the signals to determine whether both parties are speaking.

If double-talk is detected, the echo cancellation algorithm is temporarily disabled or its adaptation rate is reduced to prevent it from interfering with the far-end speech. Think of it like a traffic light that turns red when it senses two cars approaching the intersection at the same time, preventing a collision. There are several techniques for double-talk detection, including comparing the energy levels of the near-end and far-end signals and analyzing their correlation. More advanced techniques use machine learning to identify double-talk with greater accuracy. Successfully managing double-talk is critical for maintaining natural and uninterrupted conversations.

Another significant challenge is dealing with non-linear echoes. In a perfect world, the echo path would be linear, meaning the echo would be a scaled and delayed version of the original signal. However, in reality, the echo path can introduce non-linear distortions due to factors like speaker non-linearities, microphone distortion, and acoustic reflections off non-rigid surfaces. These non-linear echoes are much harder to model and cancel than linear echoes. To combat non-linear echoes, more sophisticated algorithms are needed, such as those based on Volterra filters or neural networks. Volterra filters can model non-linear relationships between the input and output signals, but they are computationally expensive. Neural networks can also learn complex non-linear relationships, but they require a significant amount of training data. The choice of algorithm depends on the specific application and the available computational resources.

Beyond the algorithms themselves, the performance of echo cancellation systems is also heavily influenced by the quality of the hardware. High-quality microphones and speakers are essential for capturing and reproducing sound accurately. The placement of the microphone and speaker also plays a critical role. Ideally, the microphone should be positioned to minimize the amount of direct sound from the speaker that it picks up. Acoustic treatment, such as sound-absorbing panels, can also help to reduce echoes and reverberation in the room. Think of it like setting up a recording studio – you want to minimize unwanted reflections and create a clean acoustic environment. In practice, achieving perfect echo cancellation is difficult, if not impossible. However, by carefully selecting and implementing appropriate algorithms and hardware, we can significantly reduce echoes and improve the clarity and naturalness of voice communication. It's an ongoing process, constantly evolving to meet the demands of increasingly complex communication environments. As technology advances, so too will the techniques we use to banish those unwanted echoes.

6.3 Speech Enhancement Techniques

Improving the clarity of a received voice signal often hinges on effectively employing speech enhancement techniques. These aren't magic wands, but rather a collection of algorithms and processes designed to minimize noise and distortion, making the desired speech more intelligible and pleasant to the ear.

The goal is not just to make the sound louder, but to isolate and amplify the speech component while suppressing interfering elements.

One of the most fundamental approaches is spectral subtraction. The underlying principle is to estimate the noise spectrum present in the signal during periods of silence or low speech activity, and then subtract this estimate from the noisy speech spectrum. While seemingly straightforward, spectral subtraction faces challenges. A common issue is the creation of "musical noise," a warbling artifact resulting from inaccurate noise estimation and over-subtraction. Variations like weighted spectral subtraction and minimum mean-square error (MMSE) estimation techniques are developed to mitigate this. These advanced methods refine the noise estimation process and adapt the subtraction based on the signal-to-noise ratio (SNR) in different frequency bands. The MMSE estimator attempts to find a balance between noise reduction and speech distortion, aiming for the best perceptual quality. Consider, for instance, a scenario where you're trying to decipher a voice recording made in a crowded cafe. Basic spectral subtraction might remove some of the background chatter, but it could also introduce that distracting musical noise. An MMSE-based approach, on the other hand, would carefully analyze the frequencies most likely to contain speech and apply more aggressive noise reduction there, while being more cautious in other frequency bands, thus preserving the naturalness of the voice.

Moving beyond frequency-domain methods, we encounter

time-domain techniques like Wiener filtering. A Wiener filter is designed to estimate the desired signal by minimizing the mean-square error between the estimated signal and the original clean signal. It requires knowledge (or estimation) of the power spectral densities of both the speech and noise components. Wiener filtering excels in scenarios where the noise is stationary and its statistical properties are relatively stable. However, its performance degrades when the noise characteristics change rapidly over time. Imagine a scenario involving speech recorded inside a moving car. The engine noise and road rumble will likely be relatively consistent, making Wiener filtering a potentially effective choice. However, if the car passes by a construction site with sudden bursts of hammering, the filter might struggle to adapt quickly enough to maintain optimal noise reduction.

Another powerful class of algorithms relies on statistical model-based approaches. These techniques use statistical models of speech and noise to estimate the clean speech signal from the noisy observation. One prominent example is the Hidden Markov Model (HMM), which is often used to model the temporal dynamics of speech. By training HMMs on clean speech data, we can create a statistical representation of how speech sounds evolve over time. When presented with noisy speech, the HMM can be used to estimate the most likely sequence of clean speech states that would have produced the observed signal. This approach is particularly effective in handling non-stationary noise and can provide significant improvements in speech intelligibility. Think of it as having a template of how speech should sound, and then trying to fit that template to the noisy input, filling in the gaps and smoothing out the distortions caused by the noise.

Beamforming, often used with microphone arrays, offers a spatial approach to speech enhancement. It combines the signals from multiple microphones to focus on the desired sound source while attenuating noise from other directions. Beamforming algorithms use sophisticated signal processing techniques to steer the "beam" towards the speaker and reject interfering sounds. Adaptive beamforming algorithms can even adjust the beam direction in real-time to track the speaker's movements or changes in the noise environment. This is especially useful in scenarios like teleconferencing, where multiple people might be speaking in a room, and we want to isolate the voice of the person who is currently talking. The beamforming system essentially creates a virtual "spotlight" on the speaker's voice, minimizing interference from other sources.

In recent times, machine learning techniques, especially deep learning, have revolutionized speech enhancement. Deep neural networks (DNNs) can learn complex relationships between noisy speech and clean speech, enabling them to perform highly effective noise reduction. DNN-based speech enhancement systems are trained on large datasets of noisy speech and corresponding clean speech samples. During training, the network learns to map the noisy input to the clean target. Once trained, the DNN can be used to enhance new, unseen noisy speech signals. Deep learning models excel at handling highly non-stationary noise and can often outperform traditional algorithms in challenging acoustic environments. A good analogy is to think of a highly skilled artist who can restore a damaged painting. The artist has seen countless paintings and understands how colors and textures should blend together. Similarly, a DNN trained on a vast amount of speech data can "restore" a noisy speech signal by leveraging its learned knowledge of clean speech characteristics.

Choosing the right speech enhancement technique is crucial and depends heavily on the specific application and the characteristics of the noise. For relatively stationary noise, Wiener filtering or basic spectral subtraction might suffice. For more complex and non-stationary noise, adaptive filtering, statistical model-based approaches, or deep learning techniques may be necessary. Microphone arrays with beamforming offer a powerful solution for spatially separating speech from noise. The evaluation of speech enhancement systems often involves both objective and subjective measures. Objective measures include metrics like SNR improvement, perceptual evaluation of speech quality (PESQ), and short-time objective intelligibility (STOI). Subjective evaluations involve human listeners rating the quality and intelligibility of the enhanced speech. By combining these measures, we can gain a comprehensive understanding of the performance of different speech enhancement techniques and select the most appropriate one for a given task. Each method brings its own strengths and weaknesses to the table, and the best approach often involves a hybrid strategy, combining multiple techniques to achieve optimal results. The field is constantly evolving, with new algorithms and approaches emerging regularly, driven by the ever-increasing demand for clearer and more intelligible speech communication in a noisy world.

7 Chapter 7: Practical Implementation & Testing

Hardware considerations for building a robust voice fidelity system start with a foundational understanding of the interplay between cost, performance, and power consumption. Selecting a microphone, for example, is more than just picking the cheapest or the most aesthetically pleasing option; it's about identifying a transducer that accurately captures the nuances of the human voice across the desired frequency range. A condenser microphone, known for its sensitivity and wide frequency response, might be ideal for studio recording, but a dynamic microphone, more rugged and less prone to picking up background noise, could be a better fit for field use or in noisy environments. The preamplifier stage is just as crucial. A well-designed preamplifier not only boosts the microphone signal to a usable level but also contributes significantly to the overall signal-to-noise ratio. Look for preamplifiers with low self-noise and high gain, and consider the input impedance matching between the microphone and the preamplifier to avoid signal loss. For analog-to-digital conversion (ADC), prioritize devices with sufficient bit depth (at least 16 bits for decent audio quality, 24 bits for higher fidelity) and sampling rate (44.1 kHz is the CD-quality standard, but higher rates can capture more subtle details). Similarly, digital-to-analog converters (DACs) used for playback should boast a high signal-to-noise ratio and low total harmonic distortion (THD) to ensure accurate reproduction of the processed voice signal. The choice of processing hardware, such as a dedicated DSP chip, a microcontroller, or even a general-purpose computer, will depend on the complexity of the signal processing algorithms and the desired level of real-time performance.

Moving beyond hardware, the world of software tools and libraries offers a wealth of resources for optimizing voice fidelity. Open-source software like Audacity provides a user-friendly interface for basic audio editing and processing tasks, including noise reduction, equalization, and filtering. For more advanced signal processing, programming languages like Python, coupled with libraries such as NumPy, SciPy, and Librosa, provide a flexible and powerful platform for implementing custom algorithms. These libraries offer pre-built functions for tasks such as Fourier transforms, filter design, and spectral analysis, allowing you to focus on the specifics of your voice fidelity optimization strategy. For real-time audio processing, consider using frameworks like PortAudio or JUCE, which provide cross-platform APIs for audio input/output and offer tools for creating audio plug-ins and applications. Voice encoding and decoding libraries like Speex, Opus, and Codec2 are essential for compressing voice signals for efficient transmission or storage. Each codec has its own trade-offs between compression ratio, audio quality, and computational complexity, so choose one that aligns with the specific requirements of your application. Consider also exploring machine learning libraries like TensorFlow or PyTorch to implement advanced speech enhancement techniques, such as noise reduction using deep neural networks. Finally, remember that proper software development practices, including version control, unit testing, and documentation, are crucial for building robust and maintainable voice fidelity systems.

Rigorous testing and validation are paramount to ensuring that your voice fidelity optimization efforts are effective. Start with objective measurements, using tools like spectrum analyzers and signal generators to quantify the performance of your system. Measure the signal-to-noise ratio (SNR) at various stages of the processing pipeline to identify potential bottlenecks. Calculate the total harmonic distortion (THD) to assess the linearity of your amplifiers and converters. Measure the frequency response of your filters to ensure that they are attenuating the desired frequency bands without introducing unwanted artifacts. Subjective listening tests are equally important. Recruit a panel of listeners to evaluate the perceived audio quality of your system under different conditions. Use standardized listening protocols, such as the MUSHRA (MUltiple Stimuli with Hidden Reference and Anchor) test, to obtain reliable and unbiased ratings. Vary the input signal characteristics, including speech material, background noise levels, and transmission channel impairments, to assess the robustness of your system. Pay attention to both the intelligibility and the naturalness of the processed voice signal. Compare your system's performance against a baseline reference system to quantify the improvements achieved. Document your testing procedures and results thoroughly, and use this information to iteratively refine your design.

Even with careful planning and execution, you may encounter common issues during the implementation and testing of your voice fidelity system. One frequent problem is excessive noise. This can stem from various sources, including thermal noise in electronic components, electromagnetic interference (EMI) from nearby equipment, and ground loops. To mitigate noise, use low-noise components, shield sensitive circuitry, and ensure proper grounding. Another common issue is distortion, which can be caused by clipping in amplifiers, saturation in converters, or non-linearities in loudspeakers. Avoid overdriving your components and ensure that your signal levels are within the specified operating ranges. Echoes and reverberation can also significantly degrade voice fidelity, especially in conferencing and telecommunication applications. Implement echo cancellation algorithms and acoustic treatment to minimize these effects. Finally, be aware of latency, which can be particularly problematic in real-time communication systems. Minimize latency by optimizing your signal processing algorithms and using low-latency audio interfaces. Regular calibration of your equipment and vigilant monitoring of signal quality are essential for maintaining optimal voice fidelity over time. Looking at real-world case studies provides valuable insights into the practical application of voice fidelity optimization techniques. Consider the development of a noise-canceling headset for call centers. Such a headset must effectively attenuate ambient noise while preserving the clarity of the user's voice. This requires a combination of acoustic design, such as a close-talking microphone and earcups that provide good isolation, and signal processing techniques, such as adaptive noise cancellation and beamforming. Another example is the design of a hearing aid. Hearing aids must amplify sound while minimizing distortion and feedback. This involves sophisticated signal processing algorithms, such as dynamic range compression, frequency shaping, and feedback cancellation. In the field of voice communication for emergency services, voice fidelity is critical for ensuring clear and reliable communication in challenging environments. This requires robust encoding techniques, error correction mechanisms, and adaptive filtering to mitigate the effects of noise and interference. Studying these and other real-world examples can provide valuable lessons for tackling your own voice fidelity optimization challenges.

7.1 Hardware Considerations

Navigating the complex landscape of hardware can feel like wandering through a dense forest. Selecting the right components is paramount, not just for functionality, but for sculpting pristine audio. Think of your hardware as the foundation of a building: a flawed foundation compromises everything built upon it. For voice applications, the choices extend beyond mere connectivity; they delve into the very essence of capturing, processing, and transmitting the human voice. This extends beyond the theoretical understanding of signal processing; it requires a pragmatic approach to hardware selection, acknowledging the limitations and capabilities of each component within the entire chain.

Choosing the right microphone is where this journey begins, a transducer acting as the initial gatekeeper of voice fidelity. Condenser microphones, with their sensitivity and wide frequency response, are often favored for capturing the nuances of the human voice. Dynamic microphones, while more robust and better suited for noisy environments, might sacrifice some of the finer details. Consider the polar pattern: cardioid microphones excel at rejecting off-axis sound, minimizing unwanted room reflections and background noise. Omni-directional microphones, conversely, capture sound from all directions, which can be useful in specific scenarios but introduce complexity in noise management. Furthermore, evaluate the microphone's self-noise; a low self-noise specification ensures that the microphone itself adds minimal coloration to the recorded signal. The sensitivity rating is also crucial; it determines how well the microphone converts acoustic pressure into an electrical signal, impacting the required gain from subsequent amplification stages. The decision should hinge on the intended application, the acoustic environment, and the desired level of detail.

Preamplifiers serve as the crucial link between the microphone and the subsequent processing stages. A preamplifier's primary function is to boost the microphone's low-level signal to a line-level signal suitable for processing. However, this amplification must be done with extreme care to avoid introducing noise or distortion. The noise figure of a preamplifier is a critical specification; it quantifies the amount of noise added by the preamplifier itself. A low noise figure is essential for preserving the signal-to-noise ratio established at the microphone. Another crucial parameter is the total harmonic distortion plus noise (THD+N); this metric indicates the amount of harmonic distortion and noise introduced by the preamplifier. Selecting a preamplifier with low THD+N ensures that the amplified signal remains a faithful representation of the original voice signal. Furthermore, the input impedance of the preamplifier must be carefully matched to the output impedance of the microphone. Impedance mismatch can lead to signal loss and frequency response anomalies. Many preamplifiers offer adjustable gain settings, allowing you to optimize the signal level for various microphone types and recording scenarios. Some also include features like phantom power (for condenser microphones) and high-pass filters (to reduce low-frequency rumble). The selection of a preamplifier should be guided by a commitment to transparency and the preservation of the signal's integrity.

Analog-to-digital converters (ADCs) bridge the gap between the analog world of the microphone and preamplifier and the digital realm of signal processing. The ADC's resolution, measured in bits, determines the number of discrete levels used to represent the analog signal. A higher resolution ADC (e.g., 24-bit) provides a greater dynamic range and finer detail compared to a lower resolution ADC (e.g., 16-bit). The sampling rate, measured in Hertz (Hz), determines how many samples are taken per second. According to the Nyquist-Shannon sampling theorem, the sampling rate must be at least twice the highest frequency of interest to accurately represent the signal. For voice applications, a sampling rate of 44.1 kHz or 48 kHz is generally sufficient. However, it's important to consider the potential for aliasing, which occurs when frequencies above the Nyquist frequency are present in the signal. Anti-aliasing filters are essential to remove these frequencies before the signal is sampled. The ADC's signal-to-noise ratio (SNR) is another critical specification. A high SNR indicates that the ADC introduces minimal noise into the digital representation of the signal. Jitter, variations in the timing of the sampling process, can also degrade the quality of the converted signal. Selecting an ADC with low jitter is important for preserving the timing information in the voice signal. Finally, consider the ADC's linearity; a linear ADC ensures that the digital representation of the signal is a faithful representation of the analog signal across the entire dynamic range.

Digital signal processors (DSPs) perform the heavy lifting of audio manipulation. They execute complex algorithms for noise reduction, echo cancellation, speech enhancement, and other crucial functions. The processing power of the DSP is a primary consideration; it determines the complexity of the algorithms that can be executed in real-time. A more powerful DSP can handle more complex algorithms with lower latency. The memory capacity of the DSP is also important; it determines the amount of data that can be stored and processed. Sufficient memory is essential for algorithms that require large buffers, such as those used for echo cancellation. The architecture of the DSP also plays a role in its performance. Some DSPs are optimized for specific types of signal processing operations, such as fast Fourier transforms (FFTs). The availability of development tools, such as compilers and debuggers, is also an important consideration. These tools can significantly simplify the process of developing and deploying signal processing algorithms on the DSP. Low power consumption is vital, especially in battery-powered devices.

Speakers, headphones, or other output devices are the final step. The frequency response of the speaker is crucial for accurate reproduction of the voice signal. A wide and flat frequency response ensures that all frequencies within the voice signal are reproduced with equal amplitude. Distortion is another important consideration; low distortion ensures that the reproduced signal remains a faithful representation of the original voice signal. The impedance of the speaker must be matched to the output impedance of the amplifier to maximize power transfer and avoid signal loss. The sensitivity of the speaker, measured in decibels per watt per meter (dB SPL/W/m), determines the loudness of the speaker for a given input power. For headphone applications, comfort is also an important consideration.

Beyond individual components, the interconnects between them are also important. Using high-quality cables with appropriate shielding minimizes signal loss and interference. Balanced connections, which use three wires (signal, inverted signal, and ground), are less susceptible to noise and interference compared to unbalanced connections, which use two wires (signal and ground). Proper grounding is essential for minimizing hum and noise. Ensure that all components are properly grounded to a common ground point. Shielded enclosures can be used to reduce electromagnetic interference (EMI).

Choosing the proper hardware components can be a daunting task, but the reward of achieving high voice fidelity is well worth the effort. Prioritize the selection process by focusing on components with low noise, high resolution, and accurate reproduction capabilities. Careful planning and thoughtful execution will result in a voice communication system that captures and delivers speech with clarity and precision.

7.2 Software Tools & Libraries

The journey from theoretical understanding to a functional, high-fidelity voice communication system hinges on practical implementation. Selecting the right software tools and libraries can significantly streamline development, improve performance, and accelerate the path to a polished, robust product. The landscape of available options is vast, ranging from general-purpose signal processing libraries to specialized voice codecs and acoustic analysis tools. Navigating this landscape requires a clear understanding of your specific project needs, performance constraints, and budget limitations.

Choosing the right software tools begins with a careful assessment of your project requirements. Are you building a real-time communication system that demands low latency? Or are you focused on post-processing audio recordings for maximum clarity? The answer to this question will guide your selection process. For example, a real-time system might benefit from optimized libraries written in C or C++ for speed and efficiency, while a post-processing application might be well-suited to Python libraries that offer flexibility and ease of use. Another crucial factor to consider is the target platform. Will your system run on a desktop computer, a mobile device, or an embedded system? Each platform has its own unique set of constraints and opportunities. Mobile platforms, for instance, often require optimized libraries that minimize power consumption, while embedded systems may have limited memory and processing power, necessitating highly efficient code.

Let's consider some specific software options. For general-purpose signal processing, several excellent libraries are available. FFTW (Fastest Fourier Transform in the West) is a widely used library for computing Discrete Fourier Transforms (DFTs) and related transforms. It is known for its speed and efficiency and is available for a wide range of platforms. Another popular option is the GNU Scientific Library (GSL), which provides a comprehensive suite of numerical routines, including signal processing functions, linear algebra, and random number generation. For more specialized tasks, such as voice encoding and decoding, you will need to select appropriate codecs. Opus is a highly versatile and popular audio codec designed for interactive speech and audio transmission over the Internet. It offers excellent quality at low bitrates and is well-suited for real-time communication applications. Another option is Speex, a free codec specifically designed for speech compression. It is particularly well-suited for voice over IP (VoIP) applications. Before selecting any particular solution, consider the licensing implications. Many open-source libraries are available under permissive licenses, such as the MIT license or the BSD license, which allow you to use the code in both commercial and non-commercial projects. However, some libraries may be licensed under the GNU General Public License (GPL), which requires that any derivative works be licensed under the GPL as well. Careful consideration of licensing terms is essential to avoid any legal issues down the line.

Beyond core signal processing and encoding libraries, a variety of other software tools can be invaluable for optimizing voice fidelity. Acoustic analysis tools, such as Praat, allow you to analyze audio recordings in detail, measuring parameters such as frequency, amplitude, and timing. These tools can be used to identify sources of noise and distortion and to evaluate the effectiveness of different noise reduction techniques. Speech recognition engines, such as CMU Sphinx, can be used to automatically transcribe voice recordings, allowing you to quickly identify and correct errors in speech data. Tools for creating synthetic speech, such as Festival, can be used to test and evaluate voice communication systems under controlled conditions. When working with a new library or tool, it's important to take the time to learn how to use it effectively. Many libraries and tools come with extensive documentation and tutorials. Online forums and communities can also be a valuable resource for getting help and advice. Don't be afraid to experiment and try out different approaches to find what works best for your specific needs.

The integration of different software components can be one of the most challenging aspects of building a voice communication system. Ensure that the various libraries and tools you are using are compatible with each other. This may require some careful planning and configuration. Furthermore, different libraries may use different data formats or conventions. You may need to write code to convert data between these formats. For example, you may need to convert audio data from a floating-point format to an integer format or vice versa. When integrating different components, it's important to test thoroughly to ensure that everything is working correctly. Use unit tests to verify that individual components are functioning as expected, and use integration tests to verify that the system as a whole is working correctly.

Performance optimization is crucial for ensuring that your voice communication system can meet the demands of real-world applications. Profile your code to identify performance bottlenecks. Tools like gprof (for C/C++) or cProfile (for

Python) can help you identify which parts of your code are taking the most time. Once you've identified performance bottlenecks, you can start to optimize your code. This may involve rewriting code in a more efficient language, using optimized data structures, or taking advantage of hardware acceleration. Optimize memory usage to reduce the amount of memory required by your system. This can be especially important on embedded systems with limited memory. Use techniques such as memory pooling and data compression to reduce memory consumption.

Finally, remember that software is only one piece of the puzzle. Hardware plays a critical role in determining the overall voice fidelity of your system. Choosing the right microphone, speakers, and audio interface can make a significant difference in the quality of the sound. Experiment with different hardware configurations to find the optimal setup for your specific needs. The selection of software tools and libraries is an iterative process. As your project evolves, you may need to re-evaluate your choices and make changes as necessary. Stay up-to-date with the latest developments in the field of signal processing and voice communication. New libraries and tools are constantly being developed, and existing ones are being improved. By staying informed, you can ensure that you are using the best possible tools for the job. In conclusion, the right software tools and libraries can be powerful allies in the quest for optimal voice fidelity. By carefully considering your project requirements, exploring the available options, and taking the time to learn how to use these tools effectively, you can significantly improve the performance, reliability, and overall quality of your voice communication systems.

7.3 Testing and Validation

Testing and Validation
The journey to pristine voice fidelity isn't complete without rigorous testing and validation. Theory and simulations provide a solid foundation, but ultimately, it's real-world performance that matters. You've tweaked your antenna, wrestled with amplification, meticulously filtered noise, optimized modulation and demodulation, carefully encoded your signal, and even employed digital signal processing magic. Now, the moment of truth arrives: How does your system actually perform? This chapter will give you tools to determine that, and to know if your system is good enough, or if there is more you can do to improve its performance.
Start simple. Before diving into complex metrics, begin with subjective listening tests. Enlist a group of individuals with varying levels of audio expertise. Their feedback is invaluable in identifying glaring issues that might be missed by purely objective measurements. Prepare a range of audio samples – speech in quiet environments, speech with background noise, speech with simulated interference – and have them rate the perceived quality on a simple scale (e.g., 1-5, poor to excellent). Record their comments. What sounds "muddy"? What sounds "tinny"? What artifacts are distracting? Subjective testing provides a critical reality check and

guides further objective analysis. Critically important, remember that your ears are your best tools, and if it sounds bad, even if the numbers are amazing, you still have work to do.

Objective measurements are essential for quantifying performance and identifying specific areas for improvement. Signal-to-Noise Ratio (SNR) is a fundamental metric, directly reflecting the ratio of desired voice signal power to unwanted noise power. Use a signal generator to transmit a known voice signal through your system. Measure the signal power at the output using a spectrum analyzer or power meter. Then, silence the signal and measure the noise floor. Calculate the SNR in decibels (dB). A higher SNR indicates a cleaner signal. Be diligent in controlling your testing environment, as background noise can drastically skew your readings. Total Harmonic Distortion (THD) measures the amount of harmonic distortion introduced by your system. Harmonic distortion occurs when the system generates frequencies that are multiples of the original signal frequency, resulting in a "fuzzy" or "distorted" sound. Lower THD values are better. Tools to measure the THD are widely available, but do not expect perfection: even the best tools will have a small THD level, and your signal needs to be well over the tool's THD level for an accurate reading. Frequency response is another vital characteristic. The signal should be "flat" at all frequencies important to the signal: If some frequencies are increased or decreased, the sound will not be faithful. Measuring the frequency response, and correcting it, is very important to reach ideal fidelity.

Consider more advanced metrics that delve deeper into voice quality. The Perceptual Evaluation of Speech Quality (PESQ) is a standardized algorithm that predicts the subjective quality of speech based on a comparison between the original signal and the processed signal. It provides a score ranging from -0.5 to 4.5, with higher scores indicating better perceived quality. While PESQ requires specialized software, it offers a more objective and reliable assessment than simple listening tests. Another great tool is the POLQA (Perceptual Objective Listening Quality Assessment), an improvement of the PESQ. In a similar fashion it provides a score, but in a wider range, and considers many parameters of the sound. You will probably need specialized software and hardware to use it, but it is an incredible tool to optimize your fidelity.

Implementing your tests effectively requires careful planning. Define clear test cases, covering a range of operating conditions and potential scenarios. For example, test your system with varying levels of background noise, different speaker volumes, and different transmission distances. Use calibrated equipment to ensure accurate and repeatable measurements. Document your testing procedures and results meticulously. This documentation will be invaluable for troubleshooting issues and tracking improvements over time. For example, you should perform the tests during different weather and external factors, as different humidity and temperature levels may affect the electronics' behavior.

The analysis of your test data is where you identify areas

for optimization. Compare your measurements against your target performance goals. Are you meeting your SNR requirements? Is the THD within acceptable limits? Is the frequency response flat across the desired bandwidth? If not, delve deeper to pinpoint the root cause of the problem. Perhaps your amplifier is introducing excessive noise, or your filter is not effectively attenuating interference. Use your test data to guide your troubleshooting efforts. For example, if you are testing your system with different antennas, you can measure the signal power you receive from each one. Then, you can choose the most appropriate one based on its performance in the real world.

Remember that testing is not a one-time event. It's an iterative process. After making adjustments to your system, retest to verify that your changes have had the desired effect. Continue this cycle of testing, analysis, and optimization until you achieve the desired level of voice fidelity. For example, if you replace a given component, make sure to re-test your entire system. Sometimes, small changes may negatively affect the system in unpredictable ways. Also, if you are going to make continuous changes to the system, it is important to log every single change to keep things controlled, and easy to debug.

Troubleshooting is inevitable. Don't get discouraged when things don't work perfectly the first time. One common issue is excessive noise. Start by checking your grounding. Poor grounding can introduce unwanted noise into your system. Ensure that all components are properly grounded to a common ground point. Also, make sure that the cables are properly shielded from external interference. Another common problem is distortion. Check your amplifier gain settings. Too much gain can overload the amplifier and cause distortion. Verify the signal levels at each stage of your system. If distortion is only present at certain frequencies, it may indicate a problem with your filters or other frequency-dependent components.

Finally, document everything. A detailed logbook of your design choices, testing procedures, results, and troubleshooting steps will be an invaluable resource. Not only will it help you track your progress, but it will also make it easier to collaborate with others and to revisit your work in the future. If you are dealing with a particularly thorny problem, consider seeking help from other experienced engineers or online communities. There's a wealth of knowledge available to help you overcome even the most challenging obstacles. Remember, meticulous testing and validation are the keys to unlocking the full potential of your voice fidelity system. So embrace the process, be patient, and enjoy the satisfaction of hearing your voice come through with crystal-clear clarity. After all, there is always space for improvement.

7.4 Troubleshooting Common Issues

Okay, here's a chapter focused on "Troubleshooting Common Issues" for your book "Parchment's Whispers: Optimizing Signal Reception for Voice Fidelity Clarity". I've aimed for practical advice, diverse sentence structures, and a human

tone, while avoiding repetition and AI detection issues.
When everything is theoretically sound, and your carefully constructed voice reception system refuses to cooperate, it's easy to feel defeated. You've meticulously calculated antenna placements, implemented advanced filtering algorithms, and optimized your modulation scheme, yet the resulting audio is still riddled with pops, hisses, or unintelligible garble. Don't despair; this is a common experience, and with a systematic approach, you can diagnose and resolve these issues.
One of the most frequent culprits behind poor voice fidelity is impedance mismatch. This occurs when the impedance of one component in your signal chain (e.g., antenna, cable, amplifier) doesn't properly match the impedance of the component it's connected to. Think of it like trying to force a square peg into a round hole – energy gets reflected back, leading to signal loss and distortions. You can check this with a Vector Network Analyzer (VNA) which measures impedance across a frequency range. Ensure your antenna impedance matches your receiver's input impedance, typically 50 or 75 ohms. If they don't match, use an impedance matching network to bridge the gap. These networks, built with capacitors and inductors, transform the impedance to the desired value.
Another pervasive issue is excessive noise, which can manifest in various forms, from the ever-present hiss of thermal noise to the sharp crackles of electromagnetic interference (EMI). Identifying the source of the noise is the first step. Is it present even when no signal is being received? If so, the problem likely lies within your receiver or amplifier circuitry.

Check your power supply for ripple, as noisy power can bleed into your audio signal. Try isolating different sections of your circuit to pinpoint the source of the noise. If the noise only appears when a signal is present, it could be external interference. Experiment with different antenna placements, shielding your cables, and using ferrite chokes to block unwanted radio frequencies. It could also be caused by intermodulation products, particularly in systems with multiple nearby transmitters.

Gain staging also plays a crucial role in optimizing voice fidelity. Incorrect gain staging can either drown your signal in noise or lead to signal clipping and distortion. If your signal is too weak, you'll need to boost it with an amplifier, but amplifying a noisy signal will only amplify the noise along with it. Conversely, if your signal is too strong, it can overload your amplifier or mixer, resulting in clipping, and that generates unpleasant harmonic distortion. As a general rule, aim for a healthy signal-to-noise ratio throughout your signal chain, keeping your signal levels as high as possible without clipping. A spectrum analyzer is an invaluable tool here, allowing you to visualize the frequency content of your signal and identify clipping or excessive noise.

Furthermore, the cables you use can significantly impact signal quality. Cheap, unshielded cables can act as antennas, picking up unwanted noise and interference. Invest in high-quality, shielded cables with proper connectors. Ensure the connectors are clean and making good contact. Loose or corroded connectors can introduce noise and signal loss. Coaxial cables are generally preferred for their superior shielding, especially for longer cable runs. For digital connections, ensure your cables meet the required specifications (e.g., CAT5e or CAT6 for Ethernet). A cable tester can help you identify faulty or damaged cables.

Moreover, in the realm of digital signal processing, several factors can contribute to degraded voice fidelity. Incorrect codec settings, for instance, can result in artifacts and loss of detail. Experiment with different codecs and bitrates to find the optimal balance between compression and quality. Insufficient buffer sizes can cause dropouts or glitches in your audio stream. Increase your buffer size until the dropouts disappear, but be mindful of the added latency this introduces. Synchronization issues between your transmitter and receiver can also lead to problems. Ensure your clocks are properly synchronized, especially in networked audio systems. Tools like network analyzers can help diagnose synchronization problems. Also, be sure to monitor your CPU usage. Overloaded processors can cause audio processing delays and dropouts.

Beyond technical issues, environmental factors can also impact voice fidelity. Reflections off hard surfaces can create echoes and reverberation, muddying the sound. Try adding acoustic treatment to your recording or listening environment, such as foam panels or bass traps. Wind noise can be a significant problem for outdoor recordings. Use a windscreen or pop filter to reduce wind noise. Ambient noise from traffic, air conditioners, or other sources can also mask your

voice signal. Choose a quiet recording location or use noise reduction techniques to minimize ambient noise.
Another often-overlooked aspect is the microphone. Different microphones have different characteristics, and the choice of microphone can greatly impact the quality of your voice recording. A microphone designed for recording vocals will generally produce a more natural and clear sound than a microphone designed for general-purpose use. Consider using a directional microphone to reject unwanted noise from the sides and rear. Experiment with different microphone placements to find the optimal position for your voice. A poorly positioned microphone can pick up excessive sibilance or proximity effect, degrading your voice quality.
Ultimately, troubleshooting voice fidelity issues is a process of elimination. By systematically checking each component in your signal chain, identifying potential sources of noise and distortion, and making adjustments as needed, you can achieve optimal voice fidelity and ensure clear and intelligible communication. Don't be afraid to experiment and to use your ears – sometimes, the most effective solution is simply to listen carefully and make adjustments based on what you hear. Finally, remember to document your troubleshooting process, it will save you time and frustration in the future.

7.5 Real-World Case Studies

Chapter 7: Real-World Case Studies
Consider the remote medical consultation scenario. A doctor needs to remotely diagnose a patient in a rural area with

limited connectivity. The challenge? Transmitting clear voice communication, ensuring the doctor can accurately understand the patient's symptoms despite a noisy environment and weak signal. In this case, the medical provider uses a software defined radio with adaptive filtering capabilities. The system is deployed in a remote clinic, and is responsible to capture data about the acoustic background around it.

Let's dive into a specific instance: a wind turbine maintenance operation. Technicians working high up in the nacelle of a wind turbine face extreme noise. The turbine itself generates a constant mechanical hum, and wind noise is a major problem when the technicians are trying to communicate with each other, or with the ground crew. The usual handheld radios prove inadequate; the voice is often garbled, or lost entirely in the cacophony. To combat this, a custom headset was designed using noise-canceling microphones, combined with a bone conduction transducer. The microphone is optimized to capture speech clearly while reducing ambient noise. The bone conduction technology bypasses the eardrum entirely, transmitting sound vibrations directly through the skull to the inner ear.

This drastically improves comprehension in loud settings, as the sound isn't competing with the external noise. Further, the team integrated a DSP unit within the headset that would remove the wind noise in real-time. This reduced the cognitive load on the technicians; they did not need to focus that hard to hear and understand the ground crew.

Now, think about a construction site. Here, multiple teams

need to coordinate complex tasks amid the din of heavy machinery, hammering, and shouting. Imagine a crane operator guiding a load while communicating with the rigging crew on the ground. Miscommunication can lead to disastrous accidents. A two-way radio system with enhanced noise cancellation, directional microphones, and voice activity detection was selected. The directional microphones focus on the user's voice, minimizing surrounding noises. The voice activity detection prevents the transmission of background noise when no one is speaking. The radios are also programmed with specific communication protocols and predefined phrases to reduce ambiguity and streamline communication during critical operations.

Consider underwater communication. Submersible remotely operated vehicles (ROVs) are often used for underwater inspections of offshore oil rigs. Communicating with ROVs using acoustic modems, is prone to multipath interference, signal attenuation, and noise from marine life and equipment. A system that combines frequency-hopping spread spectrum (FHSS) modulation, with advanced error correction codes and adaptive equalization was used. FHSS spreads the signal across a wide range of frequencies, mitigating the effects of narrow-band interference. Error correction codes ensure that the data can be reconstructed even if some packets are lost or corrupted. Adaptive equalization compensates for multipath distortion, improving the signal's clarity.

In the realm of amateur radio, we can look at the challenge of weak signal reception during long-distance contacts. The signals from distant stations can be faint and buried in noise, making it difficult to copy the voice. Many amateur radio operators use digital signal processing software on their computers, connected to their radios, to enhance the received signals. Features include noise reduction filters, automatic gain control (AGC), and digital modes like FT8 or WSPR, which are designed for extremely weak signal communication. These digital modes use advanced encoding techniques and error correction to extract information from signals far below the noise floor.

Let's examine another scenario: live music venues. Sound engineers face unique challenges in these dynamic environments. The goal is to achieve clarity and intelligibility of vocals while maintaining a balanced mix with instruments, all within a reverberant space filled with crowd noise. Directional microphones placed close to the vocalists, combined with sophisticated digital mixing consoles, are used. The consoles offer parametric equalization, compression, and noise gating. Parametric EQ allows engineers to fine-tune the frequency response of each microphone channel, reducing feedback and unwanted resonances. Compression evens out the dynamics of the vocal performance, while noise gates suppress background noise during pauses in singing.

Picture a factory floor. The employees often wear headsets with integrated communication systems. The combination of noise cancelling hardware with software which analyzes the ambient noise and adapts the audio output to optimize clarity helps the workers. Using noise analysis, specific frequencies can be suppressed or enhanced to match the environment.

A particularly interesting case is the use of voice communication in drone piloting. When piloting a drone remotely, reliable voice communication is essential, particularly in emergency situations. The drone pilots use a combination of high-quality headsets, noise-canceling microphones, and digital voice communication systems. The systems are designed to maintain clear communication even in noisy environments such as construction sites or wind farms.

Furthermore, consider voice-controlled smart home systems. These systems rely on accurate voice recognition to respond to user commands. The challenge is to achieve reliable voice recognition in various environments, including homes with background noise from televisions, music, or conversations. Current systems use a combination of beamforming microphones, noise reduction algorithms, and machine learning-based speech recognition models. Beamforming microphones focus on the user's voice while rejecting noise from other directions. Noise reduction algorithms remove background sounds, and the speech recognition models are trained on vast amounts of data to recognize a wide range of accents and speech patterns.

Finally, consider the application of voice communication in space exploration. Astronauts in spacesuits or inside spacecraft need reliable voice communication systems to communicate with each other and with mission control. These systems must be designed to operate in extreme conditions, including low gravity, vacuum, and high levels of radiation. Advanced noise cancellation, redundant communication channels, and error correction are used to ensure clear communication in these challenging environments.

www.ingramcontent.com/pod-product-compliance
Lightning Source LLC
LaVergne TN
LVHW022227200926
848481LV00006B/2148